I0487305

AQUA SOIL

Aqua Soil

Our Evolution,
Biological and Cultural

Cameron Rebigsol

A $50,000 award is offered inside this book for a solution that will remove some mathematical difficulties, such as 0 = C (speed of light), led by relativity in physics. Any college student majoring in science and engineering with one or two years of mathematical training can tackle the problem. No gimmick, no string—it is unnecessary to buy this book to win the award, but a convincing mathematical argument is hereby solicited.

Copyright © 2009 by Cameron Rebigsol.

Library of Congress Control Number:		2009911249
ISBN:	Hardcover	978-1-4415-9285-9
	Softcover	978-1-4415-9284-2

All rights reserved. No part of this book may be reproduced or transmitted
in any form or by any means, electronic or mechanical, including photocopying,
recording, or by any information storage and retrieval system,
without permission in writing from the copyright owner.

This book was printed in the United States of America.

To order additional copies of this book, contact:
Xlibris Corporation
1-888-795-4274
www.Xlibris.com
Orders@Xlibris.com
68208

Contents

PART IV   HOW WILL *RELATIVITY* SUPPORT ITSELF IN PHYSICS?

Pictures and Sources

Fig. 20-A,—B Eye Orientation Developments, by this author, for illustration purposes only

Fig. 21 **A Twenty-Four-Thousand-Year-Old Artifact,** courtesy of **Carl Zimmer,** author of *Smithsonian Intimate Guide to Human Origins,* Text © Carl Zimmer, HarperCollins Publishers Inc.

(When a Web site is referred to as a source of information, the date on which the information is found and quoted is also hereby included, for the reason that Web sites have a high frequency of coming and going.)

Preface

It is very puzzling to see that Christians and evolutionists are so determined to keep their distance from each other, but the Bible has such a strong hint that evolution is indeed a designing process of God's creation on all living beings.

Of all the living beings, humans are God's most artistic and most sophisticated creation on his land. To reach the perfection that he had designated on his aesthetical production, he also put time for the formation. As revealed in Genesis, the sequence for the creation to successively appear was first the wild, and later the human beings, not the other way around. From lifeless dust to a living being, that is what evolution is about. The only thing that seems inexplicable is the amount of time involved. The duration of time revealed in the Bible is seemingly so short that it is only a matter of a breath; but the duration of time in the evolutionists' term spans over billions of years. Do not forget that the magnitude of time between God and human beings are of completely different significance. A human has an average life span of seventy-four years, but a fly's is only about six weeks. Within these six weeks, a fly has to complete a life cycle that a human being can have seventy-four years to complete. How can a human being compare his time with God's eternalness?

Possibly the importance does not rest on how differently Christians and evolutionists have understood God's time scale, but rest on that both accept the concept that formation of perfection takes time. Every great picture begins from a simple line sketch, then rough shaping, and then coloring. The statue of David by Michelangelo started from a meaningless marble boulder. While the Christians hail and appreciate God's accomplishment, the archeologists are trying to find the disappeared lines and the evolutionists are trying to rediscover the gradual steps of how all the rough sketches have led to perfection. There should have not been conflict in between. The conflict seemingly to have appeared has something to do with the fact that we human beings all have simple minds compared to God's perfection on everything. With their simple minds, human beings have all unfortunately led themselves on a path guided by improperly exaggerated confidence—Christians and evolutionists alike.

With this exaggerated confidence, human beings believed that they either have competently understood and exposed God's secret on some issues or will be able to do so. With the simple mind and the exaggerated confidence, didn't human beings all once think that the sun revolved around the earth? When the opposite is later established, shouldn't we all appreciate God's miracle in which both the sun and the earth are so tightly connected to each other without even a rope in between? With the same history in mind, we must conclude one thing for ourselves: for the evolutionists, just because they believed that biological development in nature is seemingly governed by a certain pattern, it does not mean that they have been able to expel God from the copyright of the universe's greatest design; for the Christians, just because they feel themselves very devoted to their father, it does not mean that they can replace the creation design with a more simplified processing, such as a breath similar to a human's. However, with DNA research getting more and more pronounced, the pendulum is beginning to swing to the other side. Evolutionists advocate that life starts from a random collision of some fundamental particles, while Christians now insist that life is a creation behind which a design is far beyond human's comprehension. In short, as long as human beings cannot explain why and how the universe has come to being as it is now, we all must surrender to one Creator. Through their constant adoration toward God, the Christians incessantly remind the scientists of what a magnificent and boundless auditorium God has provided for their exploration; through their unstoppable searching in the auditorium, scientists are able to perpetually present more evidence to the Christians for them to elevate their adoration toward God.

In this book, the term "Mother Nature" is constantly mentioned. Nothing can exclude that she is one of the angels under God. God laid down rules for her operating budget, such as energy conservation, substance conservation, and chances only without naturally granted right to any living beings. He who intends to violate these rules must make himself—or somebody else—suffer. Mother Nature will not allow any dent on these rules from God.

When we refer to human physical characteristics, we must use the concept of spectrum. For example, when we mention that men have bigger body frames than women, this is from the point of view of a spectrum; we do quite often find that some individual women make some men look smaller. To relate this phenomenon with spectrum, let's draw a segment of straight line. At one end of the segment, we put a note of 100 lbs, and at the other end, 200 lbs; naturally, 150 lbs is at the midpoint. All other weights will be placed on the segment accordingly. We further use blue mark to represent all male individuals and red dot to represent all nonpregnant female individuals, assuming all individuals are of twenty years old. After sampling a large group of individuals by weight from both genders, we will find that a lot more blue dots have appeared at the end near the 200 lbs while red dots concentrated way more at the end near the 100 lbs. However, we cannot exclude that some red dots do appear near the 200 lbs' end and also few blue dots at the 100 lbs' end. From this spectrum, we say that men overall have a bigger body frame than women. Another example is that we say Caucasians have blond hair and black people have black hair, but we do sometimes find some native black people with blond hair, while some Caucasians have black hair. In other words, when a group of human beings is said to have possessed certain physical features, some individuals may be exceptions, but such an exception will not prevent the overall characteristic for this group of people from presenting a general value for discussion.

The following diagram presents an evolution tree of primates. The main trunk of this evolution has always been in the water; some branches from the main trunk extend to the land, while some branches have sunk to the water forever for some lineages to be extinct there. This main trunk has never left the water, just as a tree never leaves the soil. This is why the title of this book—*Aqua Soil*—is chosen. Evolving from some reptiles more than 100 million years ago, the primates spent their evolution life in water until the current last 1 million years, and the hominids thus emerged. Some of the hominids may have left the water life substantially earlier than the 1-million-year mark, while some may have left substantially later. In the diagram, the big green patch (refer to the back cover) represents an aqua environment, and the blue tree trunk is always in the water, with all the orange braches representing the terrestrial primates extending above the water. At the end of the tree trunk, we finally find where the human beings are, but having been isolated from water.

A lineage diagram

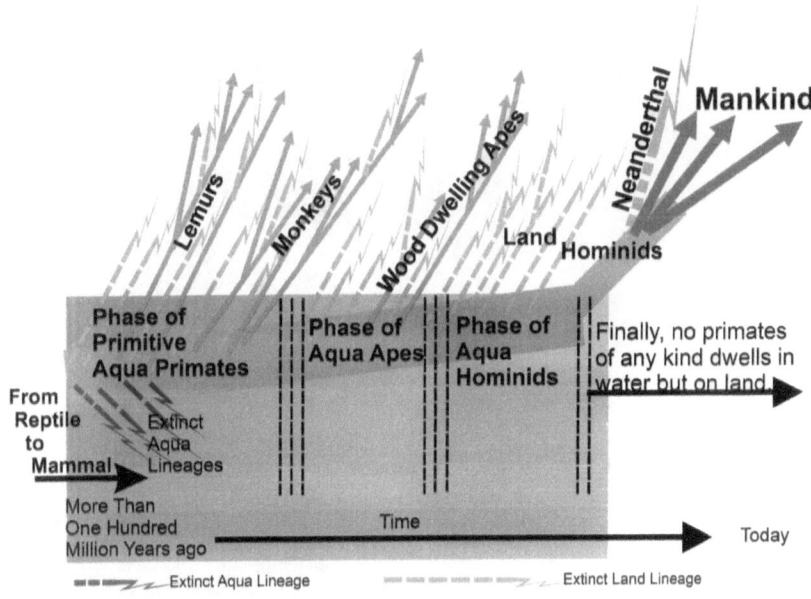

Fundamental Evolutionary Development of Primates

Part I

Biological Evolution

Chapter 1

Fundamental Characteristic
Developments

a. Primitive Aqua Primates

Modern anthropologists strongly believe that human being's ancestors came down from some trees. Evolutionists overall also believe that lives started from water. How, then, sometime in the past, did some aqua animals choose to climb up to the trees to become our ancestors? If such historical void is unable to be filled, any discussion about evolution of our ancestors naturally remains unacceptably incomplete.

In the discussion about our direct but extremely remote primate ancestors, we must prepare ourselves to deal with images of some animals that have never been mentioned with material evidence in any evolution textbook. It is for certain that these animals have long been extinct, either by sinking deep in the mud bed or lying underneath a thick dirt crust or diverging into many lineages. The most complete fossil of a primate ancestor—Ida, found in Germany—is believed to be 47 million years old; from any measure, Ida can only be regarded as a cousin of our direct ancestors. How, then, can we exclude with certainty that we even have ancestors of 60 million years old, or older, but with strong primate characteristics, such as a pair of very frontally facing eyes? But where have they been, and how did they look like?

Lacking direct evidence to show how mammals developed from reptiles, we cannot discuss in detail how our remote primate ancestors came into the picture of evolution history. However, as is discussed in more detail later, the primates' unique location of eyes alone gives us confidence to state that our direct remote ancestors had a long history of water dwelling. The water dwelling happened long before they had any chance to dwell in trees.

Not only the location of the eyes in relation to their heads, but also the way these eyes are aligned leads us to believe that only an extensively long history of aqua living can solve the puzzle. At the time they began the evolution journey as true primates, they may have been small in size. It is said that in the days that dinosaurs dominated the earth, the biggest mammals could not be more than a pound in size to have a chance to live. We may guess that our direct ancestors should have appeared on earth long before dinosaurs became extinct; so they could have been small then. It may not matter how close to historical fact this guess is, but primates still keep some version of small size nowadays. Typically, for example, a tarsier weighs less than half a pound, a pygmy marmoset is about five ounces, and a pygmy mouse lemur is even about an ounce. Figure 1 shows how small a skeleton of a tarsier looks next to a skull of a gorilla. So it may be a safe bet to assume that in those days, for a long period of time, primates were hiding themselves here and there in the size of a small frog.

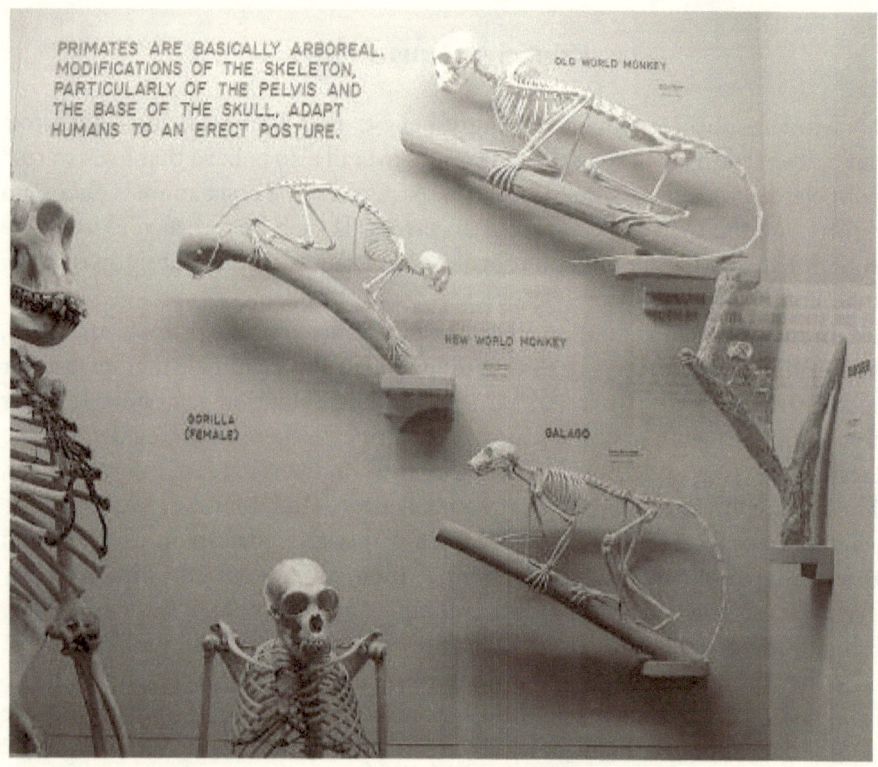

The skeleton in the far right belongs to a tarsier. See how small it is in comparison to the skull of a gorilla in the far left

Fig.1
(Courtesy of University of Michigan Exhibit Museum of Natural History)

Spending their lives in water, each of these froglike mammals must have always exposed its nostrils above water for air. But this would also expose itself to be preyed upon by predators swooping down from the sky. If some sensing organ could have provided an early warning about movement in the open air above, it certainly helped. The pair of eyes of these froglike mammals naturally took over this task.

Starting the new journey as mammals, but also as decedents of some reptiles, these animals had to carry the typical characteristics of their ancestors: their eyes were on each side of their skulls. The two distantly separated eyes presented a big problem to their visual angle. To make life even more dreadful, a bigger separation between the eyes, together with their nostrils, caused a bigger exposure of their head above the water. This certainly increased their chance to be preyed. In order to maximally hide them underwater while breathing without impairing the vision, shortening the distance between the eyes was the best choice. Gradually, more and more of those that could show this physical trait dominated the population among them; as to those that could not, a big chance for them was to have been lifted from the water as meals of the others.

With the distance between their eyes being so forcefully modified and shortened by predators form the sky, our primate ancestors were in the making. So for the purpose of illustration, let us call them *primitive primates* at this stage. Since they dwelled in water, let us further call them *primitive aqua primates*.

The gradually getting closer and closer eyes not only helped them in successfully escaping danger from the sky, but they also gave these primitive aqua primates the ability to discern spatial depth. This greatly enhanced their ability in reaching their targeted meals. Now, three elements were reinforcing in turn between each other: eyes that stayed closer increased the chance to escape danger and get more food, successful escape and the ability to acquire more meals gave them more chances of having descendants, higher chances of getting descendants helped to assure the physical traits further passing to the future.

b. An Amphibious Omnivore

Compared to other aqua mammals, these primitive aqua primates had many disadvantages. They had long limbs in relation to their body lengths. This meant that they spent more energy in the same forward thrusting force in each stroke of swimming. The frequency of strokes was also reduced because of the longer limbs. The tail they had was not equipped with a pedal-like end. They also had a comparatively wider shoulder, which created no streamline with the head, but, instead, more fluid resistance when swimming. All these came together to form a body structure that limited them from swimming like a fish or a dolphin or a seal. In short, they were not good swimmers.

—

Not only they were poor swimmers, but they also had no sharp claws either. Being tiny and without a powerful hunting tool like an eagle's talons, they could only have other small animals for food, such as shrimp, little fish, tadpoles, smaller frogs, snails, clams, small reptiles, eggs of some water fowls or reptiles, various kind of insect, etc. Judging by the common physical traits shown by various monkeys nowadays, the speculation for these primitive aqua primates to have long sharp canines should not be deviating too much from historical fact. Besides foraging for little animals for food, they also consumed vegetarian food as supplement, but animals should have been their main diet. This is so speculated because the same is observed as a common behavior with aqua mammals today. The primitive aqua primates did not grow their long sharp canines for nothing; and such physical traits have been well kept by the monkeys.

Compared to animal meals, vegetarian foods were easier to get. There was more vegetarian food readily available on land than in water. A great deal of small animals were easier to catch on land than those swimming swiftly in water too. The easier food sources thus lured the omnivore primitive aqua primates to extend their dinner table to the land—while staying adjacent to the water's edge. However, there was a price to pay for the food on land; land predators were patrolling there. So while foraging on land, the primitive aqua primates must always be prepared to rush back to water, where most of the land predators hesitate to go too far in.

Animal food provided high energy for various activities and cold resistance in winter water; vegetarian food supplemented the energy when the main diet was on shortage. With a menu of extensive choices, the primitive aqua primates found themselves fit for an extensive range of territory on earth.

c. Preparing for Bipedal Standing

Being poor swimmers, they could not catch animals in water with their mouths like a fish's or a seal's because such action demanded strong thrusts from water propelling. They must rely on their forelimbs to arrest food. This gradually degraded the strength of their mouths as a vital weapon in attacking as well as defending. Instead, the function as a weapon was gradually taken over by the forelimbs. This further introduced another physical function shifting: when their forelimbs were busy with the job of fighting or food catching, the primitive aqua primates used their hind legs to handle the job of weight supporting and locomotion.

Although living in an aquatic environment, their comparatively poor swimming ability limited their major activities to the shallow water areas. Too much danger lurked in deep water for them. It was unnecessary for them to incessantly swim in water to get air; the high content of fat resulting from the animal food enabled them to float very easily. In areas where water was shallow, the primitive aqua primates might have rested their hind limbs on the solid bottom. The water buoyancy reduced their body weight, and thus freed up their forelimbs for purposes other than weight supporting.

So beginning in an aqua environment, they gradually felt more accustomed to a bipedal gesture. When they got on land, in the absence of buoyancy from water, they must still have spread their body weight on all fours during running. However, whenever possible for activity on land, they would take advantage of their bipedal skill acquired in water too. It is true that some other mammals are also able to stand up by only their hind legs, but primates can use their forelimbs to accomplish far more complicated tasks when standing on only their hind limbs. Besides, primates showed more endurance with a bipedal gesture than other animals. The reason is that, with the help of water buoyancy, the primates' ancestors began their bipedal training in a special environment tens of millions of years ago; other animals only occasionally did it for "fun," except few such as ostrich and kangaroo.

d. Advancing Forelimbs, Degrading Swimming

The forelimbs must accomplish many tasks in the primitive aqua primates' life: locomotion, food procuring, weight supporting, attacking, and defending. Therefore, being agile and flexible would have become indispensable requirements out of these forelimbs.

The primitive aqua primates had inherited a broader chest and sideways-spreading forelimbs from their remote reptile ancestors. So stretching out their forelimbs on both sides of their body at the same time had not been a great problem for them. For more successful food accessing, their palms gradually developed in such way that the first toe and the other four toes were opposite each other. Dexterity gradually began.

As they gained more progress on the function of their forelimbs, their body structure also gradually altered itself to justify such progress. As such, the bones and muscles commanding the forelimbs near the shoulder grew stronger, making this portion of the body more massive. A more massive and more sideways-protruding shoulder could only be very negative to their swimming. To make the swimming even worse, their necks became slender in order to adjust for land activity. A slender neck was more flexible to allow the heads to turn in many angles for better hearing and visual detection.

Aside from the neck getting slender, their heads became bigger too. From the point of view of fluid mechanism, the combination of a bigger head and a broader shoulder *but* with a slender neck in between was an extremely unfavorable development for swimming. This gradually made them more and more a loser in water life. It should not be surprising that, when certain other conditions showed up, some of them gradually found that terrestrial life was an easier life.

e. Face Plane

At this point, let's introduce a term called *face plane* for further illustration. A face plane is a geometrical plane defined by the following three points on the animal's face: tip of the snout and the two eyeballs. So a face plane of a crocodile should be almost parallel to its spine; a face plane of a baboon is almost perpendicular to its spine when it stands on all its fours. Of course, a human's face plane is almost parallel to his spine too. If we draw an arrow on the face plane of a crocodile from the tip of its snout pointing toward the center between its two eyes, the arrow would point toward its rear end. Let's call it a negative plane. If, however, we do the same thing on a human's face plane, the arrow would point away from his rear end. Let's call it a positive plane. Human beings carry a face plane that is the most positive compared to other mammals.

For the primitive aqua primates, in the old time when they still carried strong characteristics from their reptile ancestors, their face planes were very negative. However, two major life requirements gradually made them adopt a body structure in which their face plane and their spine were getting more and more perpendicular to each other.

The first requirement was the manner in which they rested in shallow water: standing on hind legs while watching the sky for danger. Generation after generation of such a lifestyle would certainly bend an angle between their face plane and the spine.

The second requirement was borne out of the necessities of land activity. On land, due to the need for detecting predators and procuring food, their eyes needed to cover more angles in a three-dimensional space. In water, the eyes mainly looked upward; on land, they must also look horizontally forward, in the direction of movement. They must often look down toward the ground as well. Those who were unable to bend their head for the eyes to look forward would have a much slimmer chance of surviving on land.

f. Hair on Scalp

Being omnivores and getting food in water and land, benefiting from a broad menu, the body size of the primitive aqua primates were getting bigger and bigger each generation. Gradually, predators swooping from the sky became less of a threat to their existence. However, the diminishing threat of the predators from sky did not yield true freedom to their heads that needed to be above water most of the time. Now, their heads became attractive nutritious source for another kind of menacing protein pursuers, although not some immediate life eliminators: bloodsucking insects.

Considering the primitive primates' physical size and their manner of locomotion, venturing too far away from the waterline on land to forage might mean the price of a life. So marshland, swamps, and mangroves—either bordering at freshwater or black water—should be well-suited as their food baskets and haven for safety. However, due to the huge amount of perished and rotten substances accumulated in these areas, many kinds of insects also saw these areas as favorable quarters for infestation. As we know, many insects live on bloodsucking. When they attack, indeed, they always bombard the target in a swarm.

To deal efficiently with the swarming attack from the insects, several defending strategies were deployed by animals

1. to shield themselves with foreign materials. This kind of technique includes making underground barrows, like naked mole rats, or staying underwater, like manatees. Neither of these fitted the lifestyle of the primitive aqua primates;
2. to develop a shell, scales, or armorlike thick skin as snakes, land turtles, or pangolins. Given the physical size and the agility and flexibility that the primitive aqua primate required for survival, and being an endothermic animal, this option was absolutely impractical for them;
3. to grow hair or fur over the skin. Indeed, the thicker fur, the better. Fur was light and almost never presented any hindrance to discount agility. It is very possible that all land animals developed fur not for the purpose of better temperature regulation, but for protection against blood (or other body fluid) sucking insects. So the third technique was almost the indisputable option for the primitive aqua primates to protect their scalp, which they must expose out of water all the time. For the rest of their skin, strategy 1 had long been deployed, i.e., getting protection by being underwater.

There was also another vital function to have hair on the scalp: it protected the thin but fatty scalp from sun ray exposure.

With all characters that they had selected from Mother Nature, the primitive aqua primates were carrying the following unique features with them: a basically nude body with thicker hair on top of the scalp, two frontally staring eyes that were quite close to each other, and running on land with all fours but being able to frequently stand on only their hind legs. In addition, their ears were not long pointed like other majority of terrestrial mammals, but instead, looked like a seashell with elliptically curved outlines like many aquatic mammals we find today. Indeed, human beings' seashell-like ears should suggest a strong kinship to some aquatic mammals in the past.

Nevertheless, with all these characteristics, a group of animals that were primates enjoying an amphibious way of living with partial bipedal locomotion began to take shape on earth.

Fig. 2

Chapter 2

Different Lineages

a. Varying Habitats

The environment for the ancient primates incessantly changed. The waterline of the riverbanks and the coasts of the sea could never stay the same. In black water areas, the water levels were affected by the tides daily, and rivers could be flooded or dry up seasonally or yearly. In the long history of the earth, ocean water levels have been raised or lowered many times. All these inevitably forced the major habitats of the ancient primates to constantly shift. Such change must exert modification force on the primates to develop new features to fit the environment.

It can even be more natural to imagine that primates in different locations would have developed different features to adjust too. Even in the same big area, if some groups happened to be lured by some favorable living condition to venture deep into a thick wooded area, they may end up abandoning the original water home. If they finally succeed in dealing with the new arboreal environment, after many generations, their descendants would have looked different from the cousins that their ancestors left behind somewhere else. Groups that were trapped somewhere by natural disasters would result the same, finally bringing out a new lineage of primates.

For those that settled well in arboreal environments, an ability to grasp tree branches well was a must. The reason was obvious: they were comparatively small and had neither sharp claws nor powerful jaws with sharp teeth; staying high on trees was the most secure way to keep alive. Besides, fruits were usually high on the treetops.

To perfect the clasping ability, one toe on the foot gradually appeared opposite to the other four. In the history of evolution, different patches of primates ventured into the wood at different epochs, but they all unexceptionally developed such features on their feet. It was because they had all been modified by the same lifestyle. For those that were unable to develop such a feature, there were only two options for them: disappearing or returning to the water to join those that had not ventured too deep into the woods. No matter what time epoch, there were always some stubborn enough to find water life better. For those that continued to live in water, Mother Nature would not force them to have a foot feature like those that had dwelled in the wood. The human being's direct ancestors consistently belonged to those stubborn ones.

Bloodsucking insects swarming in the woods may not be as populous as in open areas next to the waterlines, but their bombardment was still fierce. So aside from hair on the scalp, those that settled in the woodland must also develop thick fur on the other part of the body surface for protection. For those that continued to stay in water—including humanity's direct ancestors—they had no chance and no need to develop a forklike toe split on their feet; neither did they need a thick fur over other parts of the skin besides the scalp. Water was enough to protect most of their skin against the bloodsucking insects.

Some people proposed that human beings gained a smooth body surface because their ancestors somehow had the fur shed for some reason. As far as the fur is concerned, it could provide covering for warmth, protection against bloodsucking insects, and shielding action against strong sun ray. What is the advantage for some animal to lose it? In case shedding away the fur does have some advantage for some reason that we could not figure out, why is such an advantage inapplicable to the human scalp? In particular, if the hypothesis that humans started from Africa is near factual, what makes a black skin color better than fur in protecting the skin against the sun, or to fend against bloodsucking insects, which work twenty-four hours a day, all year long? Why didn't the exchange between skin color and fur extend to gorillas or chimpanzees? If fur was a burden to animals roaming in the African savanna because of the immense heat there in the daytime, what do they have against the cold of night and winter after they shed the fur?

In case some unseen force from nature pushed human beings to extinction if fur had not been shed, why would the bloodsucking insects have left the then smooth, furless skin alone? It is not uncommon that even with thick fur like a lion's, an open wound on its body can become an invitation for the bloodsucking insects to lead the lion to death. So it is hard to imagine that fur, once possessed, would not be a valuable asset for the animal to keep in the open African Savanna but must be shed away.

Absence of a forklike structure on a human being's feet initiates another hypothesis for its explanation: they lost this feature. Did they lose it? What advantage did they gain in the process of losing it? Or did their direct ancestors ever have it?

b. Different Feet

There exists an obvious difference on the soles between the two groups of primates, i.e., human beings as one group and lemurs, monkeys, and apes as another group. There is an arch at the inner side of our soles between the ankle and the first toe, while the other group has such arch between the first toe and the second toe. It must be that both types of arches were formed after a long history of different curling actions.

The arch shape of our soles suggests to us that our direct but remote ancestors rested or moved on flatter surfaces; this arch shape is particularly good for ascending between stairlike steps. The monkeys' and apes' ancestors had an arch good for curling around rod-shaped surfaces. For the ancestors leading to the human lineage, accuracy for landing on a flat surface did not have to be as high as that of the monkeys or apes, who jumped between tree branches. Landing caused by missing a step on a flat surface was not always life-threatening, but missing a step on the rod-shaped stalks was frequently fatal to the monkeys or apes. So as long as our direct ancestors found no need to live in the woods, they felt no need to "improve" their feet structure with a forklike split between the toes. Indeed, instead of splitting the toes apart, their toes need to stick together for two important reasons: (1.) They needed a more concentrated sole area for better water propelling action during swimming. (2.) They could have a better bouncing and pawing action out of the tightly packed toes on land.

In case some of the arboreal primates did once really intend to abandon the arboreal life for any reason but to attempt roaming on the savannah, their sole structure would have brought them to extinction. The reason was very simple: when predators came, what would they do? They were not strong enough to fight back, but needed to escape as soon as possible with all fours back up high on the trees. The height of the trees played such a critical and irreplaceable role for their rescue. The arboreal primates' long and spreading toes provided weak leverage in lifting body weight for land locomotion. It was absolutely not worth trading—and no chance for the trading—to sacrifice the high speed escaping to a safe haven for bipedal movement, even a perfect bipedal movement like the humans'. Bipedal movement was a long practice of many generations, while a life could be destroyed in a split second.

It may be true that the ancestors of monkeys or apes were social animals, but the strength consolidated by their social pack still offered no match to the predators. Besides, many of the predators also formed social packs. The danger on the ground was overwhelming; there was absolutely no reason or possibility

—

for a primate to try an awkward movement on land at the price of risking a life. Although we found quite a few species of wood-dwelling primates living on grassland nowadays, we have found none of them giving up the trait of a separate first toe on their soles.

The shape of the gorilla's feet provides a strong support for the view that wood-dwelling primates were unable to attain a sole that has the absence of a forklike structure. Being outstandingly heavier than any other primates, a gorilla's feet showed the least split. With the gorilla's body size, many predators could not present threats to their existence on land, and escaping to high trees for gorillas was not an inevitable option for safety. So developing a pair of feet that must command eminent gripping power did not seem vital for their survival. Extensive ground living of 10 million years for this specie of primates should have allowed them to give up the forklike structure on their feet. However, this was not what Mother Nature had designed. The simple reason is that their toddlers, and smaller-sized females must escape to the treetops when predators approach. If Mother Nature gave no reason for the gorillas to give up the forklike structure on their feet, she would equally give no reason for another species of primate to give it up and let it develop into *Homo sapiens*, a species with a body size much smaller than a gorilla's.

Equipping their hind legs with a power not for gripping but for swimming, our direct ancestors could outmatch most of the land predators in water when escaping from land. Besides, if our direct ancestors needed to reemerge out of water to gain access to land again, it would be far more unpredictable for the land predators; an arboreal animal that has escaped to the treetops could be constantly under surveillance of the same predator. So an ape with a pair of feet that was more adaptable to water life would have a higher chance to survive the predator's chasing and will pass its physical trait to a descendant.

The possibility of the foot structure of an arboreal ape developing into ours, in case it were ever true, must mean that the first toe of our feet had been relocated from a place near the ankle to the front of the feet and became the strongest toe of the five. All arboreal primates have their feet point outward when they have a chance to stoop, while the long first toe points diagonally inward. An intense bipedal movement like the human beings' running must force their feet to point forward more to gain stronger thrust. Subsequently, the two first toes would have to turn more inward. Now, these two toes must be either in each other's way in the movement, or they could be hitting the lower portion of the other leg; high frequency of serious hurting might result from this. When this happened, Mother Nature's selection sieve first shortened the first toe to increase speed. Following this, the most time-saving arrangement for Mother Nature was to have them develop strong toes that had already been in the front of the foot, such as the second toe or the third toe. A shortened first toe and a stronger second or third toe would most likely move the first toe to an appendix position, like the dew claws of cats and dogs, or even birds. So far, we have seen no high-speed moving animal carrying a strong first toe; how do we devise an environment through which a bipedal animal can shift their first toe to the front and become the strongest?

In short, the human being's feet did not have any chance to develop from arboreal animals. To those ancient aqua primates, before they trapped themselves into adopting an arboreal life, a foot structure that had more resemblance to a human's than to a wood-dwelling primate's should make them happier in Mother Nature's selection sieve. If we accept that life did start from water, then the hypothesis of human beings coming down from the trees should not avoid convincing us why the development of our soles must go through this route: (1.) no conspicuous fork like split, as hinted by the aqua mammals; (2.) a forklike split structure between toes; (3.) to restore to a situation that a forklike split is eliminated. Did we see the same back and forth rerouting happen on other animals?

c. Aqua Apes

In the long journey of evolution, different groups of primitive aqua primates left the aquatic life behind for an arboreal life at different epochs and in different areas under different ecological conditions, yielding the broadening of the spectrum of primate species. For a long history of evolution, both aqua primates and terrestrial primates coexisted on earth. Living in utterly different environments, both primate species carried the same outstanding characteristics: a pair of very frontally facing eyes and five long toes—later called fingers—on their forelimbs. However, at some point, some aquatic primates went under an abrupt change with their body: losing the tails.

How the aquatic primates ended up losing the tails and stepping into a phase called ape is a mystery to us. However, the loss of the tail did offer some advantages in their lives. Furthermore, from the phase leading to the appearance of monkeys to the phase leading to the appearance of apes, primates seem to have had an abrupt jump in body size. A huge tooth found in Southeast Asia leads anthropologists to believe that some apes once possessed a body size eight to ten feet tall (fig. 3). This suggests to us that the loss of their tail should be strongly associated with the weight change; loss of the tail might have justified their increased size.

(Fig.3)
(Courtesy of San Diego Museum of Man)

Monkeys, great apes, and other primates all are territorial animals living in social groups. As such, war is a constant action between different individuals, families, clans, and alliances. Sometimes, these wars could be quite bloody. During these wars, a long tail could help nothing in achieving victory and instead became an expensive appendage, which just provided a vulnerable body part for the enemy for an easier control. They all became bigger and stronger now. When they became hostile to each other, any body part under an opponent's control could lead to a fatal consequence. Besides that, an extra but slim body part also served an easier target for a sneaky predator, particularly more so in water, as discussed later. When smaller in size, like their ancestors, their tails might have had provided pedaling action assisting in swimming. After the primitive aqua primates grew to a certain body size, this propelling function became more and more trivial. So keeping a tail would have nothing to gain, but everything to lose for them. In comparison, it is said that human poaching on ivory has resulted in that 30 percent of some African elephant families grow into adulthood without a tusk. So by the same token, it might be because of some brutal consequence associated with a tail, some primates who have grown into a certain size could not afford to continue to carry it.

It is well-known that animals do show preference in choosing a mating partner by a certain physical characteristics. So it was even possible that once in history the aquatic primates experienced an evolution stage in which tail carrying became a crucial factor of being rejected in getting a chance to mate. If so, then this would accelerate the process for the aqua primates to lose their tails.

With a shorter and shorter tail each generation, a new animal linage evolving from the primitive aqua primates slowly took shape on earth. But they were not termed as "apes" until their tails completely disappeared. When the first group of genuine apes appeared, they were still having their vast majority of activities in water, and therefore, these apes should be called *aqua apes*. At this epoch, arboreal apes that we found today were still animal species of a remote future. Indeed, the face feature of gorillas, chimpanzees, and orangutans together still give us strong hint about their sharing some common ancestors of water-living origin.

The tailless Gibraltar Barbary Macaques and the very short tail Japanese Macaques should give us a strong hint that there existed such an evolution period in which aqua primates prepared themselves to become aqua apes. In this period, the aqua primates kept the limb-body proportion that was closer to ours than to that of the arboreal apes. Conversely, the limb-body proportion of the arboreal apes should tell us that their characteristic physiques uniquely developed long after they isolated themselves away from water-life but on the treetops. Subsequently, arboreal apes have nothing to do with those monkeys' ancestors that had been accustomed to arboreal life; arboreal apes have their ancestors coming from somewhere else, but before they got a pair of arms that were longer than legs.

d. Nose vs. Water

Gorillas and chimpanzees and orangutans all do not have a protruding nose like us. Instead, if it is not because of the two nostrils, we could even say that their noses display no apparent shape. The opening direction of their nostrils merely presents two blowholes (fig. 4). Surrounding the gorilla's blowholes is a raised ridge of skin. Not as obviously as the gorillas', orangutans' two blowholes located on the face at the bottom of an area that is more or less concave.

Comparatively, many kinds of monkeys—such as baboon, Japanese macaque, Gibraltar monkey—carry a more pronouncing shape for their noses than the apes. Why did the apes take such a shape? The answer is the need of a constant airflow by their aqua ancestors after entering the phase of apes. When coming to the stage of ape, the aqua primates leading to the appearance of aqua apes had been living in water in a much longer history than monkeys, a history that should be measured in a scale of millions of years. With this nose feature, the aqua apes could manage to have an uninterrupted breathing easier, and the raised edge surrounding the blowholes provided a certain guarding action against water spillage from the surrounding. If they took the shape of our nose, they must either bob their heads up and down to take an on-and-off airflow or constantly exposed a bigger portion of their head above water. Such exposing had two disadvantages: to align their body more toward a vertical direction in relation to their swimming and to sink a bigger portion of their body into the water to acquire the same buoyancy. All these further hindered their swimming skill that had shown less and less favorable to their surviving.

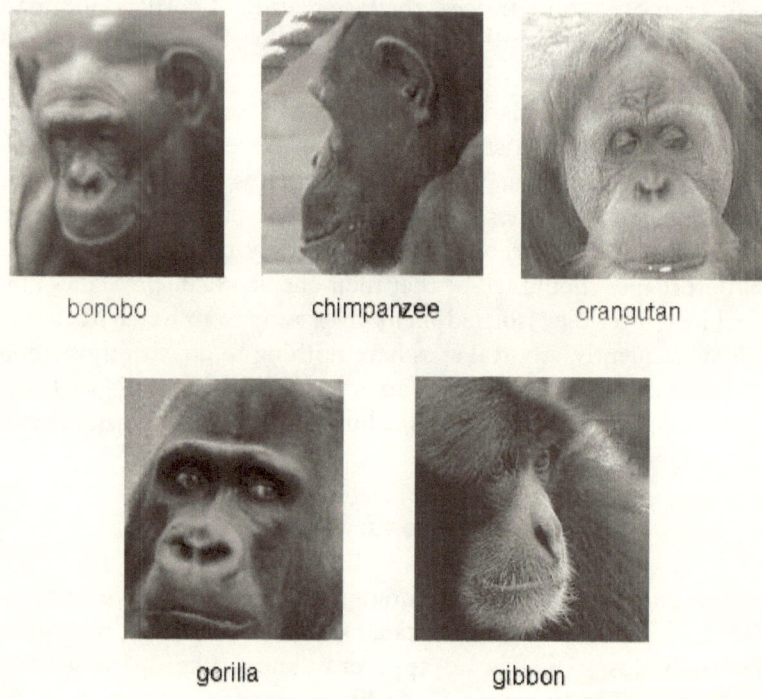

bonobo   chimpanzee   orangutan

gorilla   gibbon

Fig. 4

The shape and arrangement of their nostrils were good only for swimming in relatively calm water. To swim in vigorous water area, an aqua mammal needs a blocking mechanism for the blowholes like what dolphins or manatees have, as well as a corresponding ability to hold breath for a long diving. There was another reason for the aqua apes to shape their nose with characteristic that we detect from the arboreal apes'. We will further discuss this later.

e. Wood-Dwelling Apes

Aqua ape's body structure was not supportive to high-speed swimming. Therefore, the aqua habitat gradually presented more inconvenience to their existence. Inconvenience in the wildness meant life-threatening, of course. Terrestrial environment thus presented certain attraction to them. Fortunately, with a bigger body size and living in social pack, when venturing on land they could now defend themselves with certain efficiency against some smaller land predators. This sure gave them more confidence in accessing the easier food source supplied by plants. This led to a new lifestyle, which in turn led to the branching off of the land-dwelling apes from aqua apes. Again, as what had occurred to monkeys, different time epochs and different locations resulted in different species of arboreal apes.

After getting in wood, they encountered no obvious natural reason to force their nose to take another shape. At least airflow for breathing in wood had no difference from that in water; indeed, it was even better because water choking would no longer happen at all. Insects in dense wood were also much less swarming than near the water edges. So lacking the need for new adjustment, they basically retained the same shape of nose as when they left the water and passed down the same to their offsprings generation after generation.

Since the apes were bigger than their ancestor, the center of gravitation of the apes' body was naturally more elevated than that of their ancestors when resting on land. This higher center of gravitation proposed a new problem for the apes' balance on a tree branch in comparison to the monkeys.

If the height of the center of gravity was doubled, the corresponding body weight of an animal might get several times heavier, depending on which particular body part became more bulky. If an ape was twice as high as a monkey and assumed to be five times as heavy as a monkey, then the feet of a bigger ape needed to demand a gripping power ten times as strong as the monkeys' on the branch in order to command the same balance on the trees. This was literarily a torture to the apes' feet (fig. 5). To make up what the foot unable to handle and to overcome the fact that bigger branches that could withstand their heavier body must be farther apart, they grew with longer arms and thus created a bigger arm span. A bigger arm span had much bigger advantage than wider leg straddling between branches because it always placed the body gravity center below the pivot points, i.e., where the hands were gripping. We all know from experience that if only one supporting point is available, an object would stay with more stability when its gravitational center is located below the point where it receives supporting.

In contrast to the bulkier apes, with a smaller body, monkeys could find more single branch to support their body weight. Their lower center of gravity of the body also had less chance to tip them out of balance when they walked on a single branch with all fours. They had no need to develop a longer arm for bigger arm span. Instead, they would rather keep a pair of longer and stronger hind legs for leaping, which meant a more swift locomotion among trees. So the monkeys may well have retained the proportion between limb length and body that their primitive aqua primate ancestors had. Being closer to this proportion, our body structure should give us a strong indication that our direct ancestors had never gone up to the trees to mix with the arboreal apes; rather, our direct ancestors retained something similarly shared by the ancestors of the monkeys.

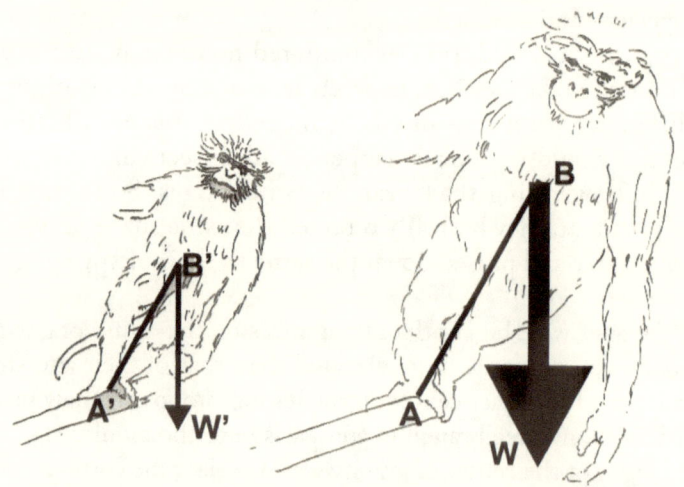

Compared to monkeys, apes' upper bodies are usually more massive in proportion. This further elevates B, their center of gravity, higher during bipedal standing besides their longer body length. If length AB=2A'B', weight W=5W', the torque produced by gravity to tip an ape out of balance on a branch would be ten times as strong as that exerted on a monkey. The ape's feet must correspondingly provide griping power ten times as strong as that for a monkey in maintaining the same balance.

Fig. 5

The bigger arm span brought two prominent body developments for the arboreal apes: First, a proportionally stronger upper body. This was a necessity for a better command on the longer arms. In contrast to the fuller development of the arms and upper body, their legs seemed to be dragging behind and underdeveloped. Second, a more obvious bipedal movement compared to any other primates except human beings. However, their bipedal movement is actually a special result of a four limb movement among tree branches. In the trees, it is their forelimbs that do the "walking" between branches; their feet just follow to give support. Since their soles and bipedal movement had been adjusted for arboreal life, when they walked on flatland, they would depend on the outer edges, other than the balls, of their soles against the ground for forward thrusting. To accomplish this, they need to sway their bodies from side to side like a pendulum pivoted upside down. In addition, a pair of comparatively weaker legs made bipedal movement on land an undesirable locomotion manner for the arboreal apes. Their heavier upper body would feel more comfortable with the support of the longer and stronger arms during movement on land. When moving on land with all fours, the longer fingers at the end of the arm could only provide poor leverage on each bounce, and their particularly heavier upper body made such leverage even more difficult. Therefore, instead of spreading all their fingers to support the body weight, they naturally curled up their knuckles to support. Knuckle walking is a more comfortable gesture for their hands, which had been so used to gripping on tree branches than spreading on ground.

If human beings were said to have evolved from arboreal apes, here comes another two related questions needed explanation besides the shape of their feet. Question 1 is whether they started coming down from trees with a body structure showing longer arms than legs or longer legs than arms. An answer of longer legs than arms would obviously reject itself because such apes, if ever existed, would extinct among trees at no time. Just imagine how a pregnant female would grip on some branches with her longer feet while her shorter arm span could not offer enough assistance. If the answer is longer arms than legs, then here comes question number 2: why must the wood-dwelling apes feel the necessity at the beginning to give up a pair of longer and stronger arms? A pair of longer and stronger arms was so vital in securing safety and so convenient in reaching for food compared to a pair of shorter and weaker arms, and such convenience was particularly important on the treetops.

Even if there was some time in history, like what some people had suggested, that food on ground appeared more abundant than on trees, giving up a longer and stronger arm for a shorter and weaker arm to face the overwhelming danger on ground was not a good trade-off. When they saw the ground presenting to them with a good food source, they would have presented themselves as good food source to too many predators as well. If it was said that they traded off longer and stronger arms for longer and stronger legs for a better running, that was an obvious very poor trade-off to start. Human being had not been found running faster with two longer legs than the apes running with all four limbs. To make it worse, the animal that was supposed to have evolved into human being even clipped off their lifeline by eliminating the fork structure on foot. When a pack of predators pursued after them, quick ascending on the tree was their only rescue, but now how could they do it without those life lines? Starting a new life with a poor or practically self-destroying strategy in an extremely unfavorable environment that they would not deserve to end up with intelligent descendants like human beings is obvious. Among the apes, gorilla is the largest. Large and strong as they are, life still demands a pair of long and strong arms to function for them besides retaining the fork structure between their toes.

Fig. 6

f. Aqua Apes' Life

To the aqua apes, until they could develop the forklike structure with their feet, wood dwelling was a lifestyle impossible to stay long. While being forced to continue to hang around in the aqua environment, retaining a pair of longer legs was required by the need of swimming. Their soles were also so justified by developing a more concentrated area for water treading.

Unfortunately, comparing to dolphins or fish, swimming became more and more a struggle for them. From time to time, their impaired ability of snatching fast-moving food in water and their incompetence against other more aggressive aqua animals got them into crisis. They must increasingly get on to land for an easier food source. But the food on land, especially the vegetation food, would supply them with less fat content than the food from water. For this reason, the aqua apes never had a chance to develop a thick fat layer underneath their skin like many other aqua mammals. The lowering content of body fat weakened their ability to resist cold temperature in water for generations to come. This further shifted their lifestyle more and more to rely on land life. But before a terrestrial life was completely settled, there was still a long process of transition, taking tens of million years.

Slowly, as each later generation showed up, their body-fat content subdued. However, compared to those primates that have dwelled on land for generations, they still retained the highest level of fat in their body. The fat they retained had made many generations of them to accustom to spending their leisure time in water by floating with bellies up. With the security feeling guaranteed by their increasing size, they could afford to spend their leisure time in this manner now, although danger may still showed up here and there if they were not careful enough (fig. 7). It is very possible in this phase of lifestyle that they lost their tails. Imagine that while sleeping with belly up but a tail was dangling in water; wouldn't it be an invitation of easy meal to some predators coming from underneath?

Life became increasingly embarrassing to them. The aqua dining room became more and more inaccessible to them with their poor swimming skill; the treetops in the heavy wood area would not allow them to thrive there because of their feet structure. So they must increasingly extend their food basket to open, dry land. The forelimbs now were more for food catching or fighting than supporting body weight for running. The proportionally shorter forelimbs made it awkward to support body weight with all fours anyway, not to mention running on all fours. Accordingly, the task of locomotion must be shifted more and more to the hind legs. The increasing land life demanded a more mature bipedal locomotion out of their body. Long toes like what wood-dwelling apes had were unjustified for efficient bipedal movement. So Mother Nature introduced to them shorter and shorter toes. However, wide contact between the soles and the ground was still necessary for the need of avoiding sinking in soft mud too easily as well as for reducing fatigue in body-weight supporting. So in compensation of the lost areas due to the shortened toes, Mother Nature extended their length and width of their soles in proportion. Those aqua apes that failed to adopt the arboreal life turned out to persistently preserve a special linage of primate on earth; but their increasing failure in water life in turn must force them to rely more intensely on land.

Fig. 7

g. Beginning of a Raised Nose

Heavier land activities were also more heavily taxed, not by ordinary predators but by the bloodsucking insects. The aqua apes that were unable to attain an arboreal life must make a choice between the following: (1) to have easier food source on land but to face more frequent attack by those insects or (2) less attack by insects in water but also less food along with their bodies' poorer and poorer fitness in water life.

The insects in open space but next to water edge were much more heavily swarming than in thick wood where their arboreal cousins dwelled. Even so, easier food source outweighed the hunger that an aqua life may frequently lead to; aqua life had been more and more hostile to them not only because of their unsupportive swimming skill but also the reduced endurance against cold temperature.

When moving to land, developing a fur to cover the entire body other than a limited area like the scalp should have been a natural choice for the aqua apes in warding off the bloodsucking insects. However, developing a fur was a long and slow process. The arboreal cousins had achieved it earlier in the evolution history. Those leaving the water must later take another period to get the same. So they must still rely heavily on the shielding action of water for the protection against the bloodsucking insects at time instant that foraging was not an immediate need or when irresistible predator appeared.

If we accepted that primates originated from water, we would find an interesting time line correlated with the fur thickness among the various species of primates. Among the primates, lemurs and monkeys, typified by gelada or mandrill, have the thickest fur, gorillas next, and then chimpanzees and bonobos. As to human beings, they practically have bare skin. This seems to reveal a certain degree of correlation between the fur thickness and the time sequence of abandoning of water life for each group.

The bloodsucking insects had forced the aqua apes to stay in water for an extensively long period during the entire journey of evolution; the water shielding was so imperative to the aqua apes. However, no matter how the aqua apes shield their body, they could not have their nose shielded. Taking the shape of blowholes as those revealed by the arboreal apes, their nostrils served as an easy entrance for the swarming insects to get inside to their body. It was even more so during intense inhaling. This was not only feeling bad but sometimes was also dangerous because the insects consistently created open wound inside the blowholes. Disease or even parasites were so directly introduced. Wasn't it nice that the blowhole could be shielded but free breath could still be kept? A raise hood over their blowholes gradually showed up. The raised hood altered the direction of the inhaled airflow; the insects found it more difficult to gain shortcut to the blowholes. Not only this, the raised hood offered an inner surface for hair to grow; this further blocked the insects from entering. As a by-product, the raised hood also helped to warm up the air intake a little; this is certainly a plus to their health.

Now, a nose—a real nose, other than just two blowholes—was taking a more and more prominent shape on the face of the aqua apes. Of course, the newly gained hood over the blowholes would further impair their swimming skill. However, because they very often spent their time in water with belly up, the advantage of a nose outweighed the disadvantage.

Now, we seemed to be able to tell that through the same time duration of evolution, a special lineage of primate had experienced more drastic change than other lines of mammals. To endure all these changes and survive the entire evolution path, such a particular species must have demonstrated some unusual biological characteristics. This unusual biological characteristic should not have been merely some simple variance or modification of body structures. Natural disasters always came at a higher speed than body adjustment. An ability to guide a rapid escape from the disaster or hardship as well as an ability to lead to, or even create, a better living environment should be far more benefiting to keep the species surviving than a limited body structure modification. This ability is the outcome of intelligence. It only came from the brain. All these meant that after surpassing all hurdles along the evolution path, a special linage of aqua apes that had intelligence began to show up on earth. With this ability, they gradually began to apply materials from the environment to compensate what their physical limitation had failed them. As this came, another kind of ape, neither living in wood nor absolutely relying on water, began to take shape in the spectrum of the animal kingdom.

In the long period of being aqua apes, by spending much of their life floating in water with belly up, they gradually molded their face plane to a positive pointing direction, closer and closer to be parallel to their spine. A positively pointing face plane parallel to the spine offered convenience for a new locomotion gesture: walking straight upright with only two hind legs while freeing two forelimbs for more complicated tasks. Indeed, even the ancestors of gorillas and chimpanzee already achieved a certain positive degree on their face plan, but they left the water way too early and missed a lot of "training."

Aqua apes, after sending gorillas and chimpanzees to ascend into the trees, with their current upright bipedal walking gesture, were grabbing stones from the ground to prepare a new and important chapter in the primate evolution history.

h. Hominids

Group after group of losers from the water life now took a new route on land with the hope that a new arena would offer them better opportunity.

Terrestrial life may be more resourceful, but not that much easier than water life. Being slowly compelled to leave water, compared to many other terrestrial animals, these ventures were very physically limited; too often they were even unable to bite to hurt other animal. They could not rely on merely their physical buildup to earn a living on land without assistance. However, besides the lifeless material objects in the surrounding, no one else could assist them. This just simply meant that they must enter a stage of applying material objects in the surrounding to assist themselves in order to colonize the non-water environment. Tool utilizing and tool making were brought to the animal world by the aqua apes that now entered a new phase of evolution. At the beginning, even a simple rock throwing but guided by a certain purpose in enhancing surviving could be viewed as tool using, and the action of simply picking up the rock for such throwing must be regarded as tool making as well. Tool using accelerated their desire as well as their chance of terrestrial living.

The transition from water life to land life took long time. Nothing was easy at the beginning: their physical unfitness in a new world, the overwhelming land predators, swarming bloodsucking insects. However, a new dwelling quarter was the only exit for continuation of living on. Those who felt more confident in their talent on tool making and tool utilizing pioneered more frequently on the exploration of land life.

Now, they slowly advanced them to a phase that is so-called hominid. During the transitional period from heavily aqua living to being completely isolated from aqua living, they should be called *aqua hominids*. It is very possible that in this phase of evolution, they had been walking with perfect bipedal gait on ground. Among them, some were trapped in situations with more frequent tool usage; this in turn kept them more distantly from aqua life. At the time they completely stayed away from water environment except regarding water as part of the food menu, they brought out the terrestrial hominids. Not all aqua hominids became terrestrial hominids at the same time at all locations. In this period, aqua hominids and many successive batches of land hominids overlapped. Eventually, after a long history of wading between water and land, the last primate that was accustomed to water disappeared from earth. The aqua primate was extinct forever. With the discovery of many fossil records that had direct indication of existence of human beings, we would like to say that one million years before today may be a time mark for the aqua hominids to disappear while some substantially earlier, such as the owner of the skeleton of *Homo ergaster* (1.5 million years), while some substantially later, such as Neanderthal (various found between four hundred thousand years to thirty-two thousand years), Peking man (seven hundred thousand years old). The lineage leading to the ancestors of the Caucasians may be even later. (Will a speculation of one hundred thousand years be too far away from fact?) On the other hand, since the skeleton of the *Homo ergaster* is so old, with the upper trunk looking so trapezoidal (fig. 18-B), nobody should be able to judge with certainty that this skeleton owner must walk perfectly straight up like modern human beings or retain a slight stooping posture when standing.

Fig. 8

i. Hair Form

When we mention hair growing on the scalp of the aqua primates, someone may argue that hair on the scalp easily blocked their vision during their swimming and thus this development should not be favored by evolution. This argument is probably drawn from the observation on modern human races with straight hair. It was not necessarily this way in the old days. In some way, the length of hair of mammals seemed being able to adjust itself. In the old time, if the aqua primates or aqua apes could grow hair that would shed at a certain length, then swimming with hair would not be a problem to them. Just like sea otters or polar bears, hair does not present a problem to their swimming. In case a gorilla or a chimpanzee could swim nowadays, the length of hair on their scalp would not interfere anything either. Another trick to avoid the hair from intervening the swimming was to develop a hairstyle that was tightly curled.

Naturally, someone may feel suspicious whether or not humans' pubic hair, besides hair on scalp, is also a result of bombardment by bloodsucking insects. How would hair needed to grow there if some aqua ancestors' bodies were always shielded by water? We notice that a human body has four smelly areas: two armpits, the crotch area, and the anus. When in the phase of aqua apes, their constant gesture of floating with bellies up always exposed the crotch areas openly. The crotch area was a strongly odorous area to attract high concentration of bloodsucking insects. Even far smellier, the anus turned out to be far less hairy. This phenomenon should give us a support of the idea that there was one time our ancestors had their crotch area exposed more frequently than the anus. Floating in water with belly up should be a reasonable speculation. Their armpits, also smelly, might not be as open as the crotch area to the insects and thus resulted in less hair growing in these areas.

We also notice that pubic hair begins to show up when humans' pubic areas start to secrete stronger odors. Is this coincidental or consequential?

Chapter 3

Origin: From Africa? Or Not!

a. The Early Primates Habitats

In those very primitive days, aqua primates were small, equipped without menacingly aggressive body part to become absolute carnivores. Avoiding being preyed became one of the essential efforts in their daily life. Like many small animals, their strategy to increase the chance of the individual surviving was to live in swarm. To guarantee such continuous existence of swarming, a high number of reproduction was necessary; their habitats must be favorable to such reproduction activity. Given their special living style and body structure, they must live in some areas that were warm most of the time in a year. Besides, water must be relatively calm, and abundant supply of fresh water was a must, even for those lived in black water areas. Surrounding these water areas, at least not too far away, must be also large areas of vegetation. So river delta areas adjacent to ocean coasts where wetland, marsh, swamp, and mangrove are overwhelmed should be the most favorable places for them to live.

If we look at the tectonic plate map for the earth about 100 million years ago (fig. 9 and fig.10), we will find that along north latitude of twenty-five degrees on the Eastern Hemisphere there was a belt of giant size islands. This belt extended from the Far East China coast all the way to the point that matched the present Gibraltar Strait in the west. Compared to any other areas in the map, this island belt should be some areas providing the most favorable living condition for the primitive aqua primates. First of all, this belt located at the best temperature zone, as it was certainly an area that was free of long and severe winter climate. The primitive primates' naked body could resist cold temperature only to a certain degree with their relatively high content of body fat, but their small size limited the energy they could store. Between the islands, water provided them easy mean for migration traffic. So as seasonal temperature varied, they could easily move toward south or north of the belt accordingly. Second, next to those giant islands in the belt, there would certainly be an abundance of small islands.

All these big and small islands would come together, embracing a big but also relatively calm water body. Compared with the nowadays geological condition, we could also be confident of saying that the overall freshwater supplies in these areas were more than ample in the old time.

Ellipse shown in picture encircles many islands that were spread in a moderate temperature zone in Eastern Hemisphere.
(Ellipse is added by this author)

Fig. 9

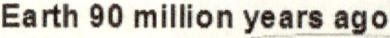
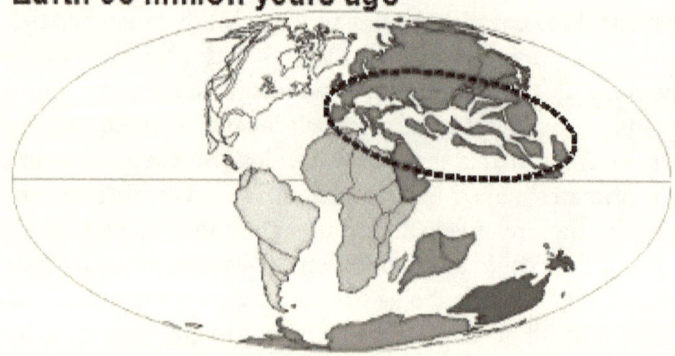

Fig. 10

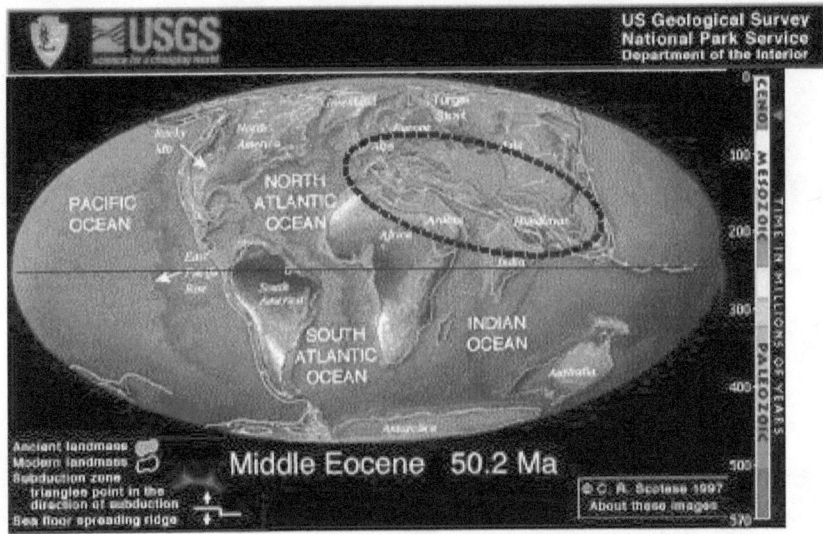

Earth 50 million years ago

Fig. 11

Between the year of 120 million years ago to now, Islands in the island belt gradually merged to each other, consolidating into bigger piece of lands. Water channels spread like a web all over the belt.

Fig. 12

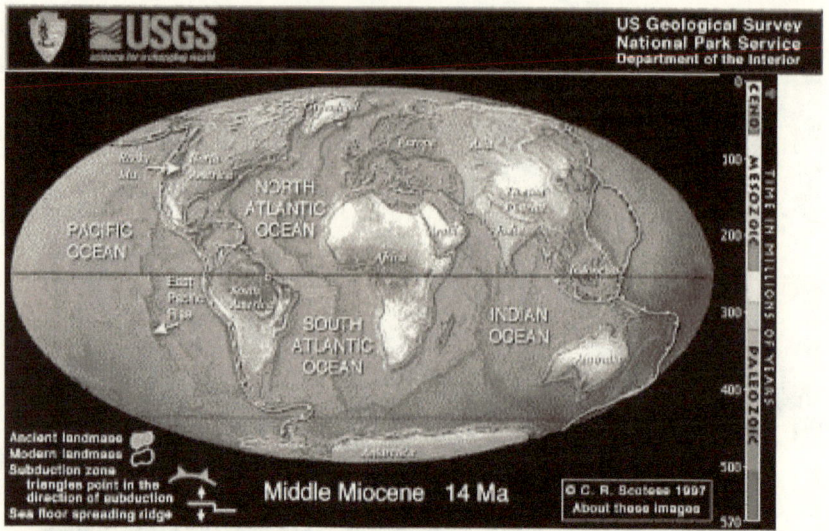

Fourteen million years ago, the Indian subcontinent began to take shape. Animal migration between the west end and the east end of the previously existing island belt can no longer be as fluidly as before. But it is still not bad, as the new land is still low in altitude and relatively flat. Although not shown in the picture, the previous water channels may not have completely disappeared but still spread all over the continent like a web of water ways but just smaller in size.

Fig. 13
(Fig. 9 to fig.13 are published by the U.S. government.)

b. The Early Surviving Strategy

The field as favorable habitats was extensively big; overall resources of water and food never ran out of supply, and temperature was comfortable. All these were favorable to high number reproduction for small animals. Subsequently countless clusters of these aqua animals spread all over in this island belt, combing back and forth on this part of earth round the year. To achieve high number of reproduction, prolific reproduction and short gestation period that was not restrained by seasons should be the most effective ways. Limited by the living condition in water, prolific reproduction could not be an option for aqua mammals. So the little primitive aqua primates chose short gestation period. It is impossible to tell what the gestation period was for an animal 100 million years ago, but we could take some data (rounded up in days) of smaller mammals for reference:

bat	40	rabbit	31	dog	61
sea otters	60	squirrel	44	mouse	21

Coming back to the primates, galago is found to have about four months of gestation period and tarsiers about six months. Some speculated forerunners of primate called Hadrocodium Wui found in Asia ("The Rise of Mammals," *National Geographic*, April 2003) that was believed to be 190 million years old had even smaller body volume than the animals listed above and thus potentially had shorter gestation period. Although they might not be the direct ancestors leading to our appearance, the point is that there could be a strong possibility that the primates leading to our existence was very tiny to begin with. If so, such primates should have short gestation period, being able to ensure the frequent replacement of the loss of individuals caused by many reasons, particularly the predators. Since food seemed to be a year-round supply near water edges, there was no reason for them to have their gestation period highly modulated by seasons. So new generations were in the making constantly.

With only one baby at a time to feed, the females of the primitive aqua primates had no need to have a body with multiple nipples like rats, pigs, or cats. Their aquatic life seemed also to have determined that their chest was the best location for the nipples. With the nipples located at where they were, a compromise between nursing and vulnerability was realized: baby could get milk and easier air access at the same time and the mother did not have to fully expose herself above water either. The way of nursing by aqua primates may have been developed in two phases during the entire aquatic period of the aqua primates.

In the early phase, baby sucked the milk while mother's nipples were in water, as mother must swim or rest in water with belly facing down. This was similar to the manner of whales nursing. But different from the whales, the primate baby must discontinue the sucking every so often and get to the water surface for air. Gradually, as the aqua primates gained in size, their nursing behavior entered another phase: a mother fed the baby floating with her belly up and the nipples were above or near the water surface (fig. 14). Babies now truly enjoyed the freedom of getting milk and air at the same time; they partially crawled upon their mothers' bodies during sucking while their lower bodies still sank in water. This phase of nursing behavior lasted an extremely long period of time in the history of primate evolution. Indeed, the female human being's perpetual swollen mammary gland should be a result of this long history nursing manner.

A more protruding nipple above water would be more convenient and more efficient for the baby's sucking. Gorillas and chimpanzees got off water way too early to retain this figure for their females. Comparatively, bonobo might have left water later than them, and the bonobo females do seem to possess a slightly swollen mammary gland, although not as conspicuously as human females (fig. 15).

The nursing-while-floating manner also potentially reinforced a gesture in influencing the future apes' physical appearance: their chins sank low to the chests. More discussion on this topic will be found in later paragraphs.

Fig. 14

Fig. 15

Nursing the baby on land in those days should not be an option for the mother and baby at all because when predators came, the baby could not cling on to the mother's slippery nude body, nor the mother could hold the baby and run quickly. Yes, evolution must eventually allow them to nurse the baby on land, but this must be waited until they gained even bigger size and, better yet, had a more advanced social behavior, such as being able to encircle the mothers and babies by a group of males. To minimize the difficulty of baby carrying by the mother, either during escaping or foraging, it was very possible in those days that baby learned how to swim shortly after birth. In comparison, in a highly unsecured environment, wildebeest baby must be able to run in a few minutes after birth.

c. More Physical Developments

The later phase of their nursing manner should have begun even before the gorillas roamed on this world. When the gorillas, chimpanzees, and orangutans have their spine in an upright position, their chins sink low toward their chest, while monkeys show much less tendency of such posture. This suggested that in the early aqua ape stage, even before gorilla or orangutan showed up, when at rest, they already felt more comfortable to float with their bellies up other than bellies facing down. When the early primitive aqua primates sent some of their comrades into woods to develop into monkeys, they had not yet got used to this kind of posture.

While their bellies facing toward the sky, the aqua primates could not help to curl their necks so that their heads could bend toward the belly more and thus be at a better angle for vision and air. This prolonged habit of so many generations left all great apes, human beings included, a bent near the shoulder blades on the spine. This bent is concaved toward the front and slightly "humping" toward the back. We do not find this bent as obvious on the monkeys' skeleton. If we compare the side view of both spines of a human beings and a gorilla (fig. 16) to the spine of the monkeys shown in fig. 1 (top two skeletons in the center), we will find a very obvious curvature on the spines of the human being and the gorilla near the shoulder blades. In order to support such bending, the apes' back portion of the neck developed some strong muscle. This characteristic is still very obvious with gorillas nowadays. The human's curvature near the shoulder blade is stronger than that of the gorilla's; this is another lead for us to believe that human beings' direct ancestors stayed in water much longer than those of the gorillas.

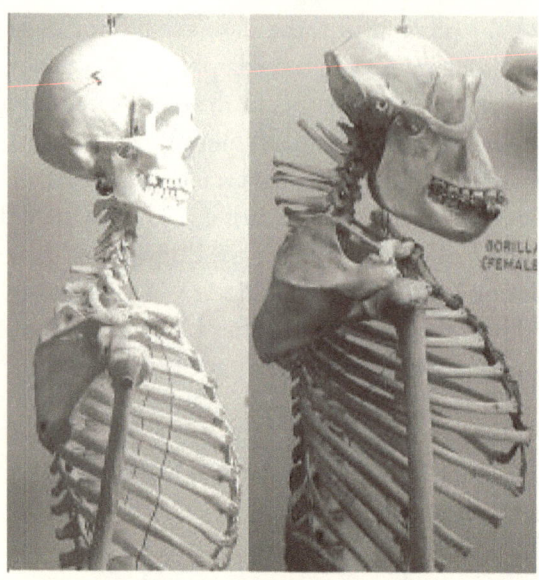

Fig. 16
(Courtesy of University of Michigan Exhibit Museum of Natural History.)

This neck bending created a problem for their swimming, as this habit reinforced a perpendicular relationship between their spine and face plane. If the face plane must be parallel to the water surface for air when swimming at a higher speed, their spine now must be in a bigger inclination toward the vertical position in relationship to the movement direction, and thus a bigger drag would be inevitably produced in their swimming.

The low-sinking chin created another problem for the aqua apes. When they moved on land, their shoulder must present blockage to their heads when they tried to move their heads from side to side. To overcome this problem, Mother Nature began to extend the length of their necks. However, then, another problem surfaced up. While floating with bellies up, the weight of their heads constituted a torque with respect to the center of floating of their entire body. This torque had a tendency of sinking their heads backward to water, prying open a bigger and bigger distance between their chin and their chest. The longer the necked extended, the bigger the torque became. This change subsequently shifted the center of gravitation of their body more toward the heads. The shift was not much but must disturb the balance with high sensitivity between the center of gravitation and the center of floating (fig.17). To reestablish the balance, they needed a torque of opposite direction to counter the disturbance. This could be done if more floating power toward the heads could be created. For this purpose, the most efficient way was to expand their air chambers, i.e., the

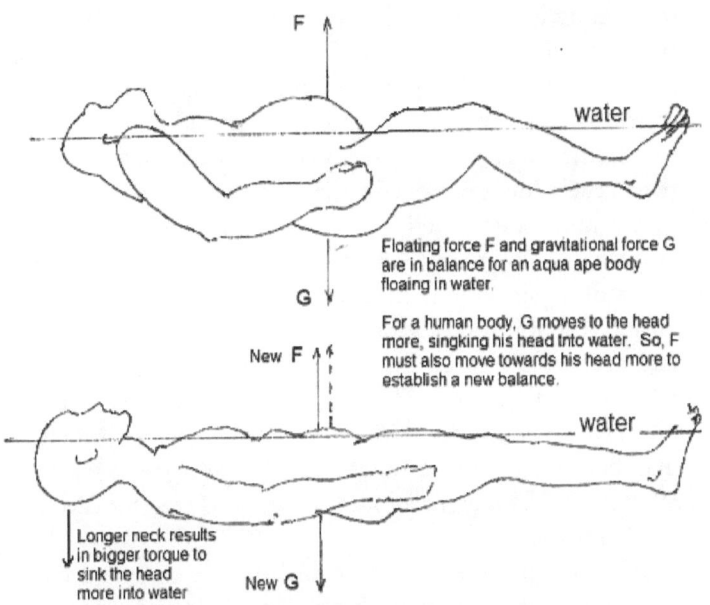

Floating force F and gravitational force G are in balance for an aqua ape body floaing in water.

For a human body, G moves to the head more, singking his head tnto water. So, F must also move towards his head more to establish a new balance.

Longer neck results in bigger torque to sink the head more into water

Fig. 17

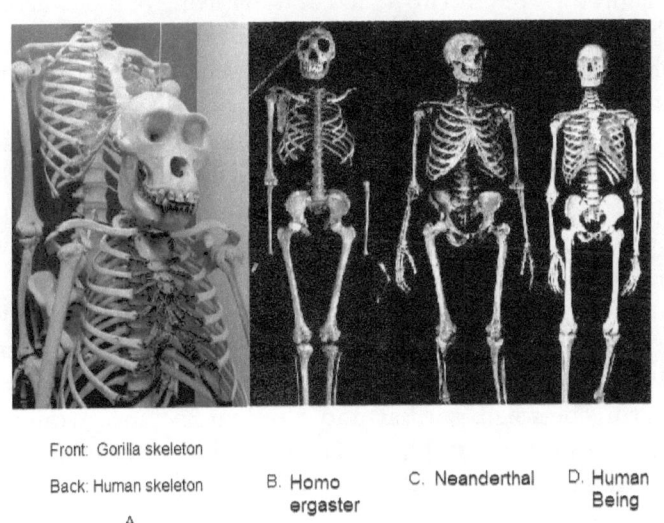

Front: Gorilla skeleton

Back: Human skeleton

A.

B. Homo ergaster

C. Neanderthal

D. Human Being

Fig. 18

A: Courtesy of University of Michigan Exhibit Museum of Natural History.

B, C, and D: Courtesy of *Smithsonian Intimate Guide to Human Origins*, by *Carl Zimmer* space where the lung was, indeed, the closer to the head, the better. Gradually, their upper trunk case gained more volume and thus a bigger horizontal dimension near the shoulder level; their belly shrunk accordingly too. When we place the skeletons of a gorilla, a 1.5-million-year-old *Homo ergaster* found in Africa, a Neanderthal and a human side by side, we could detect the very obvious change between their frameworks (fig.18).The overall tendency is that a more spacious upper trunk is developed for a more modern species of primate.

The skeleton development in such a time sequence also revealed to us that there was a time period in which even hominids could be grouped as land hominid and aqua hominid and coexisted at the same time. While some hominid ancestors like the live owner of the 1.5-million-year-old *Homo ergaster* were pioneering the exploration in the African land, a great number of other hominid ancestors still stayed in water somewhere else other than Africa to expand their upper air chamber. *It was those who extended the prolonging aqua life somewhere else were preparing the debut of the offsprings that would be called Neanderthal and* Homo sapiens *in the future.*

As the head gained more and more weight, the extended but now more delicate neck of the floating body in water was pulled to bend to curl in another direction by gravity. After the humans completely got out of water, their skeletons retained a slight trace of this late period of water living of their ancestors, i.e., a curve on the neck whose curvature was opposite to that of the bent of the spine near the shoulder blades. If we compare this between the neck of a human being and a neck of a gorilla, fig. 16 can reveal the obvious difference.

As the neck bone was developing to be longer and curving in another direction, one bone that was always unarticulated on the neck was not following. That was the hyoid. Mother Nature gave no reason for this bone to move with the development of the neck portion of the spinal column. This development gradually located (actually retained) the aqua apes' larynx lower in the throat. In addition, the slightly backward tilting neck bone made the pharynx yield more freedom to the movement of the larynx. With this new arrangement, complicated language was slowly getting on the way to the more advanced aqua apes' social life.

Comparing between gorillas and chimpanzees, we would find that chimpanzees are more vocal. That is because chimpanzees were more recent descendants of the aqua apes than gorillas and had more time in the evolution history to extend their spine of the neck portion in water and thus, in comparison, seat the hyoid lower in the neck.

d. Immigration Direction

For the sake of convenience, the term "Africans" involved in the discussion of this particular chapter is referred to those indigenous human beings found in the African heartland, or sub-Sahara. This term is used with the term "black" interchangeably. It is unfair that when the term "African Americans" is used, such term is not found to include those Americans whose ancestors have long been dwelling in the big piece of land in North Africa bordering with Mediterranean Sea for uncountable centuries.

It has been customarily circulated in the academic world of evolution that human beings were descendants of some hominids, whose ancestors were certain kind of wood-dwelling apes, coming out of Africa. The only evidence supporting such idea so far was that the majority of apes were found in Africa and that some oldest hominid fossils were also found in Africa. Continued from this out-of-Africa idea, one hypothesis suggests that a very potential reason leading to the end of Neanderthal was the appearance of the human beings that were of higher intelligence. This hypothesis has some problems. First, it advocates that intelligence played a vital role in *Homo sapiens* species replacement. In contradiction to this, such an intelligence difference is forbidden to be discovered among modern races for various reasons. However, will or will not this hypothesis predict the possibility of phasing out of some race in the future? In case that phasing out does happen, will it link to intelligence difference between races? Or more directly, will the race being less intelligent leave first? Second, ever since these Africans left Africa, have they ever been found going back to their ancestors' homeland and exerted a heavy pressure on their historical cousins' continuous existence? An answer of either yes or no will find itself difficult to conform to the out-of-Africa theory.

An answer of yes means that the current species of *Homo sapiens* in Africa is no longer the indigenous whose ancestors are commonly believed to have been rooting there for million years. Whoever they have been now, they are some species of replacement having invaded Africa no more than fifty thousand years, or one hundred thousand years the maximum, ago from today. The out-of-Africa hypothesis would have become "into Africa." Then the supposedly more intelligent African pioneers, with the ability to have pushed Neanderthal in Europe to extinct and completely replaced the descendants of Peking man in Asia, must have also removed all natives in their ancestors' continent. Then, the out-of-Africa theory must demand us to prove that the conquerors indeed first left Africa, but now came back.

Naturally, the homebound journey, if ever happened, must happen after the home-departing journey. So does it mean that the physical traits attained by modern Africans are traits that the homebound conquerors had acquired from outside of Africa? Assuming a yes for the answer, then it would become indisputable that physical traits carried by the Europeans we found today are traits developed from the African departers who left home fifty thousand to one hundred thousand years ago. This further means that the European physical trait should be more primitive; so should also be the traits carried by the Far East Asians. Logically, it must mean that the physical traits carried by the modern Africans should be the youngest of all.

Given that the Africans are said to have begun their world conquering journey fifty thousand years ago, given that the American Indians are believed to have migrated from Asia to the new land twelve thousand years ago, and given that Neanderthal are found finally disappeared thirty-two thousand years ago, thirty thousand years before today should be a fair time figure to be assumed for the home bound conquerors to complete the facial feature overhaul in Africa. If we place all human facial features side by side with that of all primates we can find diverging in these 50 million years, how should we convince us that everybody else has an older facial feature than Africans, who are said to have belonged to a version of thirty thousand years or younger? The elements to compare include the slope of the forehead, the squareness of the frontal cranium, the protrusion of the combination of the upper jaw and the lower jaw, subsequently the slope of the face plane, the size of the teeth, the size of opening of the mouth, the breadth of the nose. If any reason has existed for us to believe that detailed comparison of racial features only serves to destroy harmonious relationship between human races, a refusal of such detailed comparison should not be complained about. It is understandable. However, such refusal must immediately compromise the confidence on the acceptance of the out-of-Africa theory, for the refusal creates a corner for the theory to escape challenge. Nowadays, some people insist to advocate no racial behavior difference to be found between races based on biological reason; on the other hand, the same people are very sensitive to races in policy making but making policies obviously favorable to certain races. Possibly, some people need to show their nobility. However, nobility shown on either one side must discredit the nobility on the other side. It is certainly an ideal political harvest to be able to reap credit of nobility on both sides; why shouldn't academic nobility surrender?

An answer of no to the question of homebound conquering journey will raise this question: why did these African pioneers find themselves so incompetent in conquering Africa while they were able to conquer everywhere else in the world? There is no evidence showing that lands outside of Africa were devoid of bipedal primates of high intelligence who were capable of toolmaking and long-term planning. If Europe cannot yet produce evidence to prove the existence of an ancient ancestor for Caucasians but Neanderthal there has only been replaced there, at least, someone must place Peking man in a proper historical place. Were the descendants of this five-hundred-thousand—to seven-hundred-thousand-year-old man completely replaced?

Did the out-of-Africa pioneers love anywhere else but exactly lose interest in taking over a land that has been believed to be so good as to have given rise to the most advanced version of primate? Were the indigenous even more able than these coming back pioneers so that all the homebound invasions were stopped? Were the pioneers that the Africans sent out to conquer the world only the less-able individuals? If so, why did the home-staying but even more able African indigenous suddenly stop the egression history? Such suddenness happened beyond comprehension and was illogical. Although assumed less-able than the home-staying cousins, the supposed pioneers who left Africa fifty thousand years ago obviously found some successful living somewhere else. Their success must have lured an uninterrupted emigration flux out of Africa. Who would refuse to follow the paths leading to success? Or, being failure to detect and follow the successful path of the pioneering cousins, the home staying ones must be viewed as belonging to some less able collection. So, the home staying collection must have double "personality": Being more capable and less capable than the pioneers at the same time. Logic obviously demands another route to substantiate the "out-of-Africa" assertion. So far, humans have only found a history that complains that Africa was violently intimidated by outsiders. In those years that the only rule to be respected was the rule of the jungle, it would be extremely puzzling to learn that those who were supposed to be more able by heritage were intimidated by the descendants of the ones of less success. If the immigration flux has never stopped, the more-able species must have continued what they have done to Neanderthal: to wipe out the local habitants. Then naturally, local habitants in Europe would not have been allowed enough time to develop to become Caucasoids, neither the local habitants in Asia had time to develop into becoming Mongoloids. Instead, people nowadays would have carried a much more uniform facial feature all over the world.

Has anyone ever found any historical evidence to support the out-of-Africa immigration route? It is not to the knowledge of this author. On the contrary, in later discussion, evidence disproving this out-of-Africa idea seems quite standing out by itself. The later DNA research seems to join to support the out-of-Africa idea because the same basic DNA structures had been found from people both inside and outside of Africa, with the African indigenous carrying more diversified versions. The problem is that this DNA research still fails to evidence the immigration direction. What could it do to remove a claim made by someone saying that African apes and African people were colonists from a world outside of Africa? Indeed, some DNA tests have shown that the modern version of human beings may well have embedded with genes from ancestors that had been residing Asia for more than 2 million years before today. ("Lovers, Not Fighters?" by John Whitfield, *Scientific American* magazine, March 2008) Besides, process of DNA sampling can be dominated by biased ideas. One theory claims that, according to some DNA sampling calculation, Caucasians' blue eyes started from one girl in some Ukraine tribe about twelve thousand years ago. The theory holder may want to reconsider if he is informed of that some cats are found to carry blue eyes, that indri-indri, one species of lemurs, also carry blue eyes although in lighter shade, and that a high percentage of Japanese Macaques carry greenish eyes. In other words, human's blue eyes may well have started ever since there were primates. What is safe to bet is that cultural selection does reinforce the blue eyes in the European area to become a dominant physique among human beings there.

With all the aforementioned discussion, including the analysis of the shape of the human feet, their naked bodies but hairy scalp, and the length proportion between their body and limbs, and the squareness of their upper trunk, we could well point the finger at the other direction for ancient hominids' immigration: wave after wave, some primitive people got into the jungle of Africa from other continents. No local African apes that we have seen or not seen can develop to have carried a pair of human feet, a hairless skin except the scalp, and a body of outstandingly higher fat content than any other apes living there. African environment has offered them no chance to do so. It can be ascertained that, in any epoch in history, any living being with a human body, particularly a pair of human feet, but an ape's brain would extinct in the African jungle at no time.

The direct ancestor of all *Homo sapiens*—i.e., both African and non-African people, including Neanderthals, Peking man, Java man, *Homo erectus*—stayed in water until a very late epoch. Between aqua apes and aqua hominids, there was no sharp time line to distinguish them but a long period of mixing transition. There could have been many lineages of them that stopped existing in between. To make it short, aqua apes first produced a linage of wood-dwelling apes and later on, after another long period of aqua dwelling, gave rise to some species that are called hominids, which included an aqua version and a terrestrial version. However, many early terrestrial versions may have all gone extinct; only the last aqua version thrived and finally handed down some offsprings that are called *Homo sapiens*.

There is another argument standing out to disagree with the out-of-Africa idea. Investigation shows that the average brain size of African people is about one hundred cubic centimeters smaller than people outside of Africa ("Race, Brain Size, and IQ," *http://www.apa.org/divisions/div1/news/summer2002/rushtonpdf.pdf*, November 15, 2007). This article is not in a position to discuss the relationship between brain size and IQ but to insist a question: why Africa, a land that is said to have initiated the development of human brains, must turn out to limit the development of human brains? It is not even an accidental happening but has been historically so because the long gone-Neanderthals even had brains three hundred cubic centimeters larger than that of African people.

Furthermore, the out-of-Africa idea has been repetitively formulating the following immigrating pattern in self-supporting: Some hominids, *Homo sapiens* included, got out of Africa and earned prosperity in land outside of Africa but must later be wiped out. Then again, Africa would supply another batch of human ancestors who came out of a land where they could not thrive, then without exception, found better living in the outside world, and again inevitably got wiped out later. Then Africa supplied another batch . . . Have we ever asked why Africa, if said to be the cradle of human origin, had been so consistently hostile to human development but was able to never exhaust her source of supply of human ancestors? Or why lands away from Africa, such as Europe and Asia, are always lands enabling the hominid to thrive then must become a death trap for the entire species?

While big-brain ancient hominid in Africa is absent from the archeologist list, small-brain ancient hominids were found outside of Africa, such as the 1.7-million-year-old fossils found near Dmanisi, Republic of Georgia ("New Find," by Rick Gore, *National Geographic* magazine, August 2002). With these fossils, scholars have brought in an idea that hominid was able to migrate out of Africa with a brain smaller than that was assumed to be required. Why must we "conclude" that the live owners of these fossils have walked out of Africa? Is there any evidence against that they could have been living beings walking over here from somewhere else? Or why the appearance of small-brain hominids here must be a result of migration of a long journey? What would not permit that some hominids, beginning from their ancestors with even smaller brain, had long been dwelling around Dmanisi and gradually thrived? Their remote ancestors might have been immigrants to this local, but why must be from Africa?

The only answer to all these questions is some kind of biased assumption. However, Mother Nature does not go along with this bias. Together with the discovery of Ida in Germany, Mother Nature has told us with more than enough evidence that she did sow the seed of primate in an extensive area; but it is the human beings themselves "concluded" that Africa was the only land that permitted the primate to evolve into *Homo sapiens* for some unknown reason. Evidence is that no ancient hominids of big brain are found in Africa, at least not yet, but both small-brain and large-brain hominids have been found outside of Africa. If we must accept versatilities of DNA samples as evidence of African origin, how should we reject the evidence of versatilities of brain size of various hominids?

During the long history of primates, there was no doubt that some species of primates had migrated out of Africa and spread out to other part of the world, but the migrating flux of the other direction should have been far more abundant. With the discovery of the so-called Lucy fossils, the out-of-Africa idea seems reinforced. However, just like DNA, the most Lucy can do is to affirm that she and us as hominids shared many biological characteristics in common, but unable to point out the migration direction.

—

e. About Lucy

Lucy was found in the Afar Triangle in the North African horn. This small area was encircled by mountains of over three thousand meters high except on the north by Red Sea. Historically this was a volcano-active area. The entire area has flatland only near the Red Sea, and this flatland has no adequate freshwater supply besides the small Awash River that terminated in a salt lake called Abbe. The poor geographical condition there could hardly establish an ecological environment to start and sustain a long-term existence of animals as unique species that had Lucy's size, not to mention to allow this species to thrive and prosperously spread. So the origin of Lucy's ancestor could not have locally developed but could only have come from two sources: (1) from the heartland of Africa or (2) from a land across the Red Sea or Gulf of Aden. Comparing these two sources, source 1 was far less likely to happen than source 2.

The mountain areas that encircling the Afar Triangle could easily become a death trap if some apes walked into it with bare skin or a fur of thickness no better than the chimpanzees'. In those high-altitude areas, snowing might not be uncommon. The thick fur of geladas, the longtime monkey residents in this mountain area, should give us a strong indication how harsh the local temperature might sometimes get to. For somebody like Lucy, without sharp claws and long canines, unable to run quick because of bipedal locomotion, catching animals for food was an extremely difficult task. They should be extremely lucky that they had a chance to hunt but not to be hunted. Vegetation was lean and thin in this area. After enjoying a relatively wealthy life for generations in the heartland of Africa, Lucy could not withstand the absolute poverty here.

The historical tectonic map should tell us that 3 million years ago, at the age of Lucy, Red Sea was far narrower than nowadays. Any animal group moving from the Asian side to the African side would have found it relatively easy to cross the Red Sea. Along the Red Sea coast on the Africa side, the Afar Triangle was the only big open lowland. So chance would give animals much higher probability to get to this area from the other side of the sea than from the other side of the mountain. Particularly, primates were descendants of aqua animals. After they crossed the sea, they could not move farther into the African heartland from here, as the living condition in the mountains beyond the triangle was bleak. Possibly close to be barren, at least this lowland could support them for a while with limited freshwater supply when their population was not too big. However, any comparatively dramatic natural disaster such as a longtime drought or an epidemic might easily wipe them out.

It was possible that new batches of aqua apes carrying newer gene kept coming across the Red Sea, but it was even more possible that patches after patches of aqua ape perished here after a certain length of dwelling because of the poor living condition. Finally, some toolmaking hominids arrived. However, the local condition still put a limit for them to thrive here. After many generations, their descendant also found that Red Sea was wider and the path of returning to the old land of many generations ago was getting more and more difficult to access. Sadly, they were trapped in a land of no exit. In some of those days, Lucy inevitably died. Because this land was not a land that could inspire heavy human activity, her fossil, along with some tools from other periods, then had less chance to be disturbed and even may be preserved by volcano ash and be discovered.

f. A Jewelry Box of Gene Copies, Not a Cradle

Africa has a big portion of its continent in a mild temperature zone bordering next to the island belt. It has never stopped receiving batches of batches of immigrants of aqua primates at various epochs in large scale. However, her generosity gradually stopped when Sahara slowly took shape. When the immigrants came, the most convenient spots for them to land on Africa would be the following: (1) along the east coast facing the Indian Ocean and (2) the north coast facing the future Mediterranean Sea. Both coasts had wide open lowlands. The lowlands on the east coast were furnished with plenty of freshwater source. The north coast, however, did not seem so optimistic with freshwater except areas near the Nile River. It should not be surprising that immigrants of various kinds of primates incessantly wandered in and out of this continent, but the influx should be lot more than the efflux. The difference between the influx and the efflux was caused by the African geographical topography.

Beyond the aforementioned lowlands, the African continent had a series of highlands and mountains on the east side of this continent, the so-called African Rift System. This system separates these lowlands and the vast African heartland. The highlands and mountains also showed some infrequent low opening here and there, or even some big water bodies on the highland, bridging the coastal lowlands and the vast plains in the African heartland (fig. 19). In these plains, freshwater supply was all over. With enough freshwater and in a favorable temperature zone, a plain naturally produced plenty of food for all kinds of animals. Once some primates along the coasts infiltrated into these plains through the various corridors, they would find these plains as paradises. It might be easy for the pioneers to be attracted to get into the heartlands, but difficult for their descendants to get out. Both the comfort of the concurrent life and the difficulty of egress due to the topography bolted them tight here. So after many generations of selection by Mother Nature, these descendants gradually developed their own genes.

Fig. 19

It could be imaginable that primates in the African heartland should not have too much opportunity to exchange gene with the outside world. It was not only because they were so remotely located from the coast, but also because they were so "locked up" by the mountain barriers and the Sahara. With the islands slowly drifted northward to the future Indian subcontinent, emigration out of Africa was getting more and more impossible. The most active gene exchange "bedroom" should be on the north coast where it faced the future Mediterranean Sea. After Sahara eventually took shape, even this bedroom shrank. Besides the Nile corridor, a big part of the previous bedroom became a death trap for primates coming in and out of the African heartland. If it was not for the Nile corridor, gene exchange between the African heartland and the outside world had practically stopped. After Sahara desert took shape, the frequent gene exchange between the African continent and the outside world was basically limited to a narrow strip along the North African coast next to Mediterranean Sea. However, the majority of primates on the narrow strip that took part in the gene-exchange activity were no longer directly from the African heartland.

It is said that Sahara began to show up about one million years ago and finally settled in about five hundred thousand years before today. It is to say that the extensive gene exchange between primates of African group and non-African group could have lasted to a period that hominids appeared quite humanized. Around this time epoch, it seemed that Neanderthals began to roam in Europe and the Peking man showed up in Far East Asian. With the "door" tightly shut by the formation of Sahara, gene copies of living beings of any kind in Africa, except migrating birds, had essentially zero chance to be contaminated by the outside world. High degree of biological "purity" was thus kept. If hominids there were descendants of immigrants from the island belt, genes found in this continent should contain abundant information about the outside world but of the early days.

Historically, with the highly developed brain of hominids, no animal of any kind could have killed hominids in a larger scale than hominids themselves. It is very possible that there might once existed quite a few groups of hominids that carried different versions of DNA roaming on earth; at least we have aqua version of hominids and land version of hominids. It may also be possible that after numerous events of bloody genocide cleansing, only the group that carried the type of DNA that Lucy happened to carry has been left roaming on lands outside of Africa. Some of the hominid descendants of other versions of DNA carriers got "locked up" in the heartland of African continent. They were lucky enough not to join the genocidal-cleansing events outside of Africa. This leads to the nowadays DNA researchers to conclude that African people show more DNA varieties than people outside of Africa. The reason that more DNA varieties are found among African people is not because Africa is the cradle of all human beings, but because Africa is a jewelry box where more DNA versions have been preserved.

People from the northern edge of African continent and people from the African heartland appear strikingly different. In comparison, covering the same distance of about four thousand miles on the Far East Asian area, the facial feature between people from northern China and the majority people from Indonesia look a lot less different. Sahara therefore must be responsible for the dramatic different look between people in Northern Africa and the African heartland.

g. Ending of the Island Belt

When the island belt finally settles in the way it looks today, a big water body is formed at the west end of the original belt, presenting to us as Mediterranean Sea, with Black Sea as a less significant attachment. The majority of the bigger islands closer to the East Asian end of the belt gradually consolidate to form part of the biggest continent in the world. Weather condition around the Mediterranean Sea and Black Sea are usually far less violent than that in the original east end of the belt, i.e., East Asia, especially Southeast Asia. As far as the Mediterranean Sea is concerned, the northern shore area must have been strikingly more favorable to hominids than the southern shore. On the north shore, the freshwater supply is ample, while on the southern shore, the freshwater is trickle if Nile were not there.

South East Asia in some way still keeps the characteristic of the original belt: islands after islands are embraced by some big water body. However, the water body here is too open, and the water power as well as the weather condition is far more violent than in Mediterranean Sea area. Dwelling in this area, hominids must historically deal with year-round devastating events caused by unexpected natural power. With the calmer climate around Mediterranean Sea, hominids dwelling in this area historically need to spend far less time to deal with natural disasters. Generation after generation, the weather difference between the east end and west end of the original island belt brought out hominids of strikingly different look. Together with the difference on physical appearance is the difference on the civilization. At the west end, Mother Nature has allowed more opportunity for hominids to accumulate experience and wealth with far less interruption than hominids at the east end.

As far as physical appearance is concerned, if we put gorilla, chimpanzee, Caucasians, and blacks (i.e., all apes around the Mediterranean Sea) as one group and orangutan, gibbon, Northeast and Southeast Asians (i.e., all apes near the East Asia) as another group, we can tell that each group owns its unique physical characteristics. The apes near the Mediterranean Sea usually have heavier eyebrows, while the apes near the East Asia have smaller eyes and flatter faces. Overall, the body sizes of the apes at the west end are bigger than those at the east end. At some later paragraphs, we can discuss how the weather should have been responsible to these differences. At least, however, at this point, the physical differences but with local uniqueness can put a serious doubt on the African origin of human beings. Obviously, local uniqueness of apes has long been established. Although different lineages later diverged locally again, each of them could not get away from some common biological heritage that had been locally "carved." Even Neanderthals, a hominid species in Europe that some scholars decline to accept as *Homo sapiens*, show more resemblance to Caucasians than to Asians on their facial reconstruction work done according to their unearthed skeletons.

Chapter 4

Human Races

a. Local Uniqueness

As the consolidation of land mass in the island belt was in progress and the formation of Asia subcontinent was getting more and more prominent, the waterways between islands for the aqua primates to freely move around slowly disappeared. Subsequently, the fluidity of gene swapping among the aqua apes, or later the hominids, in this belt faced more and more restriction. In the old days, so long as the water was not too violent, they would not have to be too worried about being drowned during their aqua voyage for distant "marriage" because their fatty bodies could always float very well. With the diminishing of these water channels, restriction on the voyage made hominids gradually show up with more and more local uniqueness; more varieties of hominids diverged. In the west of the original belt around the Mediterranean Sea, hominids led to the appearance of Mediterranean hominids, which later led to the black and whites *Homo sapiens* while at the other end of the original belt appeared the Asians, typically represented by Mongoloids.

On the facial feature and body structure, more similarity can be found between whites and quite many groups of blacks (not all) than would between whites and Asians, or between Asians and blacks.

First of all, the average blacks and whites are more massive than Asians, and then we also find most whites as well as a big percentage of blacks having deeper eye sockets than Asians. Both the blacks and whites have multiple eyelids with longer eyelashes, but the folds on the Asians' eyelids as well as the eyelashes are less noticeable. The overall nose volume of the blacks and whites are much bigger than that of Asians. The hair of blacks and whites are curly (although the blacks' hair is more tightly curled) or wavy; hair shaft is finer, and their children's hair are more obvious to have this characteristic. The Far East Asians' hair, on the other hand, is always straight and with thicker hair shaft. Furthermore, both blacks and whites have better lactose tolerance than Asians. (Taking with reasonable doubt, some claims this to be a characteristic developed in only about the recent ten thousand years.) Some data even showed that blacks and whites as one group have higher rate of twin birth than that of Far East Asians. This indicates that black women and white women as one group produce more eggs, and thus their reproductive organs function with more similarities. The females can afford to have more eggs only because the natural environment is more favorable to reproduction. This means that historically local natural disaster in the Mediterranean area has been less frequent or less hostile in magnitude to their effort of rearing young. Climate difference between the west and the east locations therefore must have had a big share of responsibility in formulating the difference between reproduction systems of these two groups of people.

All these characteristics can be so uniquely and locally identifiable to these two groups of people and may be attributed due to three facts: (1) The remotely ancient ancestors of one group might have an earlier complete isolation from aqua life than the other. (2) Even if they all quitted the aqua life at about the same epoch, one group's aqua life was interrupted more frequently for a land life for some reason, such as violent weather interference, thus accumulating less water "credit" in their entire evolution history of aqua living. (3) There is a historical difference between the diets of these two groups of people ever since their remote ancestors.

b. Face

Some paragraphs back, we mentioned about the unique blowhole like nostrils for the arboreal apes. They were descendants of some aqua apes that needed to shape their nostrils this way for a continuous airflow in water. However, not only the blowholes served as a direct entrance for the insets to invade their body, but they also introduced water choking very easily by locating at a low spot of a concaved area on the face. To overcome both inconveniences, a hood over the nostrils gradually showed up, providing some shielding action over the blowholes. To start, the ancestral aqua apes leading to the appearance of gorillas developed a raised skin ridge surrounding the nostrils. But after they entered the wood and brought out the descendant of gorillas, or chimpanzees, such skin ridge lost the reason to continue to rise. In contrast, those ancestral aqua apes that did not get into the woods but continued to stay in water needed incessant improvement on the water-choking problem. On the one hand, Mother Nature forced them to grow a taller and taller hood to shield over the blowhole; besides, the ancestral aqua ape also felt the need to eliminate the concave surrounding the nose, which served as a funnel directing water to their blowhole. Slowly, the bone structure around the nostrils elevated.

Probably we have noticed that orangutans have the least noticeable skin ridge but a more obvious concave shape surrounding their nose than the other wood-dwelling apes. This can serve as a hint for a speculation how violent weather in the Southeast Asia may have affected the physical appearance of the overall apes there. The violent weather agitated violent water power. To keep themselves alive, the apes living there must escape to land when storm overpowered their endurance. Once on land, they stopped getting "training" with lessons from water life. Later, they may returned to water, but they then also have accumulated some credits from the terrestrial "classroom," and subsequently less credits from the aqua classroom. The more terrestrial credit with less aqua credit has resulted in a less noticeable skin ridge and also a less elevated bone structure around the orangutans' nose. This climate "tradition" also exerted a similar effect in formulating the facial traits of the Far East Asians. Overall, Far East Asians look flatter on their faces. This is one indication that their remote ancestors were chronically interrupted from water life for some reason. Because of a less raising bone structure surrounding the nostrils, it provides less space, or shorter dentition arc, for the teeth alignment under the upper jaw, resulting in the incisors with a somewhat shoveling look in many of them. On a full face portrait without smile, aligning the bottom of the nose and the bottom of the ear on the same level, many of them may be seen with a comparatively larger proportion of nostrils.

The water power agitated by local climate in Far East Asia was frequently devastated and even threatened life year-round. When the aqua apes in this area had developed enough intelligence to deal with the land life, they would prefer risking their life in water as little as possible. In other words, not only Mother Nature but also their intelligence reinforced more terrestrial credit in their course of evolution.

Besides the uniquely raising nose, the shape of foot and the hairless skin but a hairy scalp, we *Homo sapiens* carry more characteristics to tell us that we have been more remotely separated from the wood-dwelling apes than we thought. One of the characteristics is the color of our lips. Our lips have a distinct color from the skin surrounding them, while the wood-dwelling apes' lips do not. Another outstanding characteristic is the cleft all human beings have between their upper lip and the nose. This is an absolutely functionless characteristic. We might not be able to figure out why Mother Nature chiseled this dent on our face. As useless as it is for surviving, however, no wood-dwelling apes carry it, but all of us human beings must have one. This cleft may be a leftover of some ancestral animals that have close kinship to other mammals, such as dog, camel, and rabbit. Thus, this cleft even brings out one imagination: the aqua apes that finally brought out the appearance of *Home sapiens* were very stubborn. To any cousin that they found without this cleft, they chased them away from the water environment.

c. Body Size

To further support the reasoning that Asian aqua apes left water comparatively earlier, or were interrupted more frequently by terrestrial life, is that Asian people show less fat content in their flesh than the Mediterranean group, blacks, and whites alike. Calmer water power in a milder climate zone and a little lower overall temperature in the Mediterranean water system made it less urgent for the aqua apes there to leave the water life. But this was not what the Asian aqua apes encountered. Furthermore, higher deposit of fat in the body definitely benefited to a more sustaining period of water staying for the Mediterranean aqua apes. Before the aqua apes effectively used land material to cover their body, a lot of times staying in water made them feel warmer than in the wind-chilled air on land. We human beings no doubt have this kind of experience during swimming in a cold day too. As long as the food supply from water was still reasonably easy, they would prefer very much to stay in an environment that their ancestors had been accustomed to for so many generations. In reasoning how human being females have developed a pair of perpetually swollen mammary glands, the difference of average size of the mammary gland between the Mediterranean females and the Asian females can conversely reveal that the Asian females have received more terrestrial credit than the Mediterranean aqua apes in their entire evolution journey.

Besides the factor of water credit accumulation, the body-size difference between the Mediterranean group and the Asian group might well be molded by the historical difference on diets too.

A calmer weather condition allowed aqua primates in the Mediterranean region to procure higher percentage of food from the ocean water. In comparison, the violent water activity forced the Asian version of aqua primate to have less activity in the ocean water but more on land, particularly the mainland. More time on land naturally meant more intake of terrestrial food. As a rule of thumb, the marine animals as food are more nutritious with higher fat content. Not only the Asian group may have had more terrestrial animal as food, but they may also have higher percentage of vegetation in their menu. After so many generations on different diets, the body size between the Mediterranean group and the Asian group of hominids must display the difference thus resulted. To explain the body-size difference between people in different ecological zones, some theory proposes that the ratio between body volume and the body surface determines their body size through natural selection. This explanation has its point, but it fails to cover the following groups of people: (1) Polynesians, particularly Samoans, having large body size, they live in places straddling across the tropical zone but surrounded by ocean; straddling across the same zone, but not surrounded by ocean, Asians as well as American Indians do not look as large. (2) Eskimos, not very tall overall, live in the same temperature zone as many tall and big indigenous of other genetic origin. (3) Tutsis, average close to two meters tall for male, are found living in Africa near the Equator; on the same hot land, African pigmies are also found.

---

d. Eyes

Accumulation of aquatic credits also leads to a difference on the eye arrangement between the Mediterranean group and the Far East Asian group of people.

Let us draw three lines on a human's face plane: one line across each of the human's eye along its long axial, one line that is central to the face along the nose. We will find that the two lines across the eyes would form a T with the line that is along the nose for the Mediterranean group of people. Comparatively, a more or less of a Y shape would be noticed among a certain population in the Far East Asian group of people. This T and Y difference is another evidence for us to believe the critical influence of water life in the evolution history among our direct ancestors. Let us start this explanation with the eye arrangement shown in fig. 20 for some reptiles.

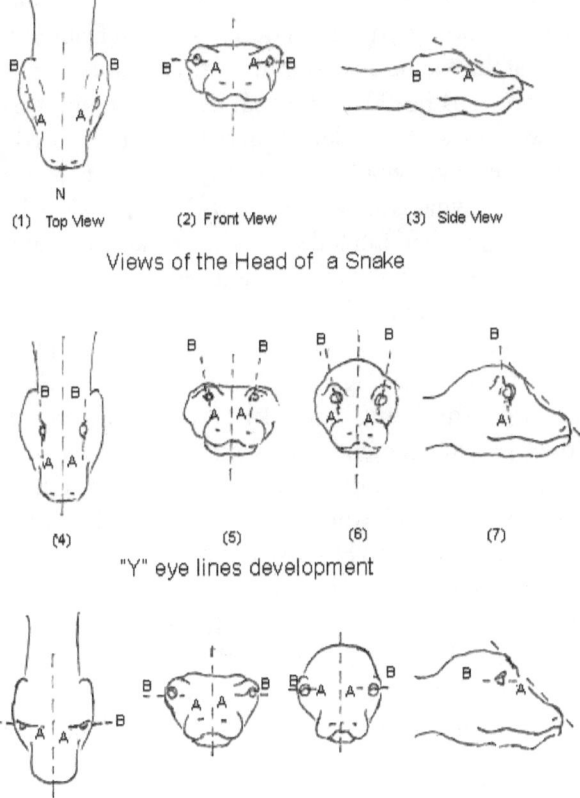

(1) Top View   (2) Front View   (3) Side View

Views of the Head of a Snake

(4)   (5)   (6)   (7)

"Y" eye lines development

(8)   (9)   (10)   (11)

"T" eye lines development

Fig. 20-A

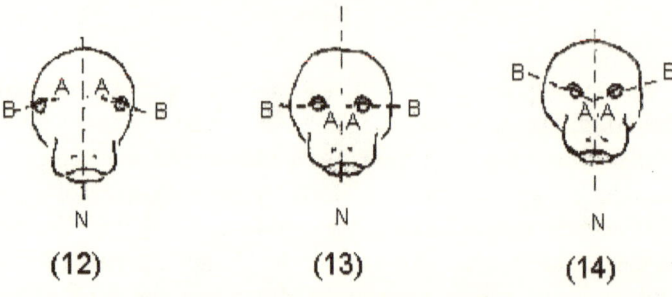

Fig. 20-B

Typically, a line drawn across a snake's eye along its long axial direction AB can be regarded as almost parallel to its central face line, N (fig. 20-A-1, and—3) When looking at the front view of fig.20-A-2, we can imagine point B of the eye line AB to be behind the page but point A to be in front of the page. At the very early history of the evolution of our primate ancestors, the feature should be similar to those shown by fig. 20-A-1,-2,-3. Beginning with this feature, if they only needed to watch the sky to detect predators and exposed the nose above the water for air, their eyes would have to be "squeezed" by water toward the center of the skull, like what is shown in fig.20-A-4 and—5. A big percentage of crocodiles and alligators we found today possess this kind of features. Continuing from fig. 20-A-4,-5, as the dome of the cranium of an animal gradually rises, a facial feature depicted by fig.20-A-6 and—7 and fig.20-B-14 gradually shows more and more prominence. The end result of all this is that a Y arrangement of eye orientation is slowly taking shape.

However, water "squeezing" and sky watching are not the only forces to influence the eye arrangement. Chewing and prey pursuing are also two dominant forces in determining the eye arrangement. Chewing requires a stronger jawbone. A stronger jawbone gradually pushed point B of the eye line AB away from the nose line N. However, chewing alone would not complete the job in eye orientation. In pursuing prey at high speed, animals need to focus on the target in front of them. As such, to shift their eyes more toward the front of the face became justifying. Chewing and the highly target-focusing together resulted in a face feature that is depicted in fig.20-A-8 (top view) and—9 (front view). T arrangement on the eye orientation was taking shape in this process. The face feature of fig.20-A-8 and—9 is very typical among many marine mammals, such as seals and sea lions, which must rely on their eyes in pursuing preys. To other kind of marine mammals, such as dolphins and killer whales, they do not develop such face feature because they rely on sonar other than their eyes in pursuing preys. For eye development taking this route, fig.20-A-10 (front view) and—11 (side view) together show that as the dome of the cranium rises, the T arrangement of the eye orientation is not affected much. As a matter of fact, if the front part of the cranium along the central line N develops relatively more rapidly in the process of eye relocation, a slightly inverted Y arrangement may even result in, as shown in fig.20-B-12.

To summarize, the eye arrangement for mammals, reptiles as well, on their face feature is historically determined by five natural forces:

1. sky watching
2. water squeeze, the need of minimum exposure of the skull above water when breathing
3. chewing
4. prey pursuing
5. rising of the dome of the cranium

Depending on the lifestyle in the long evolution history of each individual species, each of these forces has a different weight in determining the eye arrangement of that particular species, resulting in a "racing" between the T feature and the Y feature. All these five natural forces also exerted different pressure on animals at different locations at different epochs. To most of them, forces 1 and 2 may have been a long gone history but may still persist in affecting some, such as crocodile. It is said that only one in one hundred baby crocodiles can grow into adulthood, and quite a number of them are consumed by birds. So retaining the eyes on top of the skull would sure help them to lower the risk of being caught by predators from above, and thus a relatively safer childhood can be preserved. Forces 3 and 4, particularly 4, result in the dramatic difference on the eye arrangement between herbivore mammals and carnivore mammals.

To the carnivore mammals, such as lions, leopards, and hyenas, their eyes are very frontally facing and also form a very obvious Y arrangement with the nose axial line. To the herbivores, their eyes are basically on each side of the skull and lack the frontal facing feeling. It can be logical to say that the ancestors of both of these two groups of animals left the water life in a very early stage of evolution. While force 1 and force 2 had not had enough time to squeeze their eyes to come very close to each other, they already began their land life. To the herbivores, the group left water even earlier; they were always the targets of carnivores that chased at high speed from behind. So a pair of frontally facing eyes would definitely serve them for no good purpose. Instead, they would rather to have a pair of eyes that could constantly glance at what happened behind. Therefore, force 4 gradually affected their eye arrangement in a way opposite to what happened to carnivores: retaining their eyes on each side of the skull, a feature more resembling to the reptile ancestors.

As to all primates, they should have completed the shift of their eyes to the front of the face at a very early stage in their evolution history. As analyzed above, only carnivores need to have a pair of frontally facing eyes. Therefore, our remote ancestors must have begun their life as carnivores, or at least to have animals as their main diet. (Now, should we still feel puzzled to see that so many species of monkeys having long sharp canine teeth although their main diet being vegetation?)

Not every primate could wait for such eye arrangement to be completed, though. lemurs were typically such an "impatient" group. Although their eyes are very frontally facing, they do not come very close together to each other. The lemurs abandoned the aqua life and ventured in wood earlier than many other species of primates. They subsequently received less training on sky watching and therefore had less opportunity to have their eyes squeezed together by water.

Primates' unique eye characteristics all over on earth must suggest that it became possible only if gene exchange between them had been as freely as it could get, regardless how they may have been geographically separated. This further suggested that there must have existed some easy facilities to have conveyed all these "grooms" and "brides" together. Naturally the webbing water channels in the ancient island belt would be the sole candidate for such "wedding" vehicles in question. It was because of the high efficiency in traffic provided by water commuting that primates' typical eye arrangement had long been stabilized. Later on they gradually diversified into many different species in broad areas, but they must carry what their common ancestors passed on to them.

Now, it should be easier for us to understand that eye arrangement between all human beings carries so much resemblance, but local distinction is also hard to be negligible. The Y eye orientation was a result of more interruption on water life for groups of a certain locality. When their dome of cranium was developed comparatively more in advance in time than the other four natural forces had done, an orientation closer to the Y than T showed up. On the other hand, because of the fluidity in gene exchange in the old days, T and Y characteristics mixed as vigorously as they could get. The result of the thorough mixture on T and Y is that we could not find any group of people whose eyes could look as prominently in a Y shape as lions. Interestingly, the lions' Y arrangement of eyes should suggest to us that there might have existed a long history in which their ancestors also needed to worry about predators coming from the sky before they became the kings of the jungle.

If the Y eye arrangement was a result of a relatively more advanced pace of cranium dome rising, the T eye arrangement then could mean a speedier frontal shifting of someone's eye-opening. For the ancient primates of the Mediterranean group, this speedier shifting created a problem: The bone between their eyes was getting narrower and thus becoming weaker faster. This weakened bone and the two comparatively bigger hollow eye sockets came together constituting a fragile and thus vulnerable weak line across the middle of the face bone. However, the portion of the bone above this line held a big volume of pastelike material, the brain. When in rapid movement, the brain material would exert a strong impulse momentum against the wall of the cranium. This could be dangerous unless the cranium received a better support. A way to solve this problem was to thicken the part of the bone between the eyes, the nasal bone. The only way to thicken the already narrowed bone is to have it raise more above the face plane. As such, the rising of nose was gradually becoming more obvious. This structure of the nasal bone can be compared to a slanted lateral support underneath a horizontal shelf that is attached to a wall. In comparison, the Far East Asian group felt less need to reinforce this part of bone because of (1) the slower eye-opening rotation toward the front of the face at the early stage of the evolution, (2) the Y line arrangement playing a role to offset the weakened line across the two eye openings, and (3) smaller eye openings.

There stands another eye characteristic to tell the existence of the ancient migration along all those water corridors of the island belt in the old time. In the southern part of Far East Asia, we would find higher percentage of people with more obvious double eyelid folds than in the north, and the eye openings are bigger for more of the southern people too. Multiple eyelid folds are quite characterized with the people of the Mediterranean group.

This is not to say that primates in the southern part of Far East Asia inherit such characteristic from the Mediterranean primates. It was possible that multiple eyelid folds were initial characteristics for the remotely ancient ancestors of all primates. From the map of the ancient tectonic plate, we can tell that in the old days the southern ancestors in the Far East Asia and the ancestors in the west island belt could enjoy higher degree of convenience of intermigration. Comparatively, ancestors in the northern part of Far East Asia would be more highly restricted from such intermigration with the West (fig. 11, 12, 13). The migration and the subsequent gene exchange between southern Far East Asia and the Mediterranean primates continuously reinforced the eye characteristic on each other between those ancestors. In the northern part of East Asian, climate there was drier, colder, and wind-chill sandstorm covering big area was quite often; the ancestors there must squint or close their eyes for eye protection. Serious situation like this was not yet the worst. In a big open area that Gobi desert was only part of it, it snowed. Lacking trees to moderate the snow glare under the sun, the open snowfield could cause temporary blindness to people ever since their ancestors. A pair of big eyes in such environment certainly reduced one's awareness of the approaching predator. In other words, yearlong environment forced human beings' ancestors there to carry smaller eyes with thicker eyelids; and the multiple fold for their eyelids thus became less noticeable. Due to the highly restricted gene exchange in the northern part of Euro-Asia land, the single eyelid fold and the multiple eyelid fold characteristic could not reinforce on each other with the same chance like in the southern part of Euro-Asia. Of course, gene exchange must also happen between the northern and southern ancestors in the Far East Asia. The result is this: in the southern part of Far East Asia, the average peoples' eyes show something in between in such a spectrum—medium size of eye opening, with noticeable but not outstandingly conspicuous double eyelid folds, less prominent Y orientation.

e. Nose

Aqua life may render us an explanation why the monkeys in the New World had their nostrils spread wider apart than their Old World counterparts.

As a body characteristics handed downed by reptile, mammals' nostrils were located quite far apart from each other on their face to begin with. The prolonged history of sky watching not only had pulled their eyes closer, but had also squeezed their nostrils closer together. In contrast, many amphibious animals we see today, such as crocodile, hippos, frogs, not only their bulge eyes are still far apart, but so are their nostrils. Still carrying this feature of widely separated nostrils, some of the ancient primates migrated to the nowadays New World when the South American and African continents were still neighbors in close distance. The migration might not be too difficult for them if they happened to migrate in the all-time low of Atlantic Ocean and there were some midway islands in between for their fatty and highly floatable body to take a rest during the journey. The "rest" may even last several generations on those islands

Gradually, the South American continent drifted farther and farther away; whatever the characteristics they carried to the New World must only continue to develop according to new ecological environment. Weather in the New World could sometimes be quite devastating, particularly in areas near the ocean. On the other hand, thick woods offered them good protection. Therefore, it was very possible that the New World monkeys completely isolated themselves from the water life quite a time ahead of the Old World counterparts. However, the land life exerted no critical force to change the shape of their nose. So even after many species had diverged from the first few batches of immigrants, their offsprings retained the more widely separated nostrils. In the Old World, lemurs are also found to have two nostrils open sideways. This would easily lead us to relate the lemurs and the New World monkeys with some common ancestors, although the New World monkeys should be a later version of primates than lemurs. After the New World monkeys parted away, the Old World monkeys continued to indulge themselves in an environment of better living condition in water. The price to continue to stay in water was the extensive sky watching, which in turn continued to squeeze their nostrils together.

Among the Mediterranean group of *Homo sapiens*, the ancestors leading to some blacks abandoned the water life earlier than that of whites. Indeed, the similarities between the whites and blacks mentioned a few paragraphs back do not cover all blacks. It is quite noticeable that quite a high percentage of black people have a broader nose. The earlier isolation from water life gave the remote ancestors of the blacks less time to pull their nostrils together. Their hood over the two nostrils had to follow this geometrical requirement accordingly. As to the ancestors of the whites, longer aqua life enforced their nostrils to pull closer, as well as elevating the bone structure around the nose. If we assume the ancestors of the whites in about the same physical size of the blacks, then, they both need the same amount of oxygen to sustain the body activity. In order to have the same intake of oxygen, the whites need a more raised and subsequently more pointedly looking hood over the nostrils that have been closer to each other.

f. Settling for Land Life

Geographically, an utmost bigger portion of blacks would have ancestors who had closer affinity to Mediterranean Sea than to Pacific Ocean. The map of tectonic plate development showed that, ever since 14 million years ago, the earth had a long period of time in which an extensively big water body was enclosed by the land of the future Europe, Asia, and Africa. With such special environment, a great diversity of aqua apes would have been given rise around such water body and be "sent" to various areas by Mother Nature. As far as comfort was concerned, Africa should be the first choice for these apes because of its milder temperature and the abundant supply of food (before Sahara took shape, though) and the closeness in distance. This Mediterranean water cradle continued to last for a long time even after land hominids first appeared in this world. Before Sahara settled, hominids between African heartland and Mediterranean region should have close blood relationship because of constant seasonal migration among them. On the other hand, hominids from Far East Asia should be too far away for African hominids to become relatives. Indian Ocean has always been too large to cross without modern vessel.

g. A Face That Disagrees

Seasonal migrations of hominids between the African heartland and the Mediterranean region, regardless how popular and how intense they once were, must come to stop after Sahara finally took shape. A twenty-four-thousand-year-old artifact of a woman's head unearthed in France (fig.21) seemed further to suggest that large scale of immigration, either in or out of Africa, had not been happening for a long time period. The feature shown by this artifact is virtually the same looking of the contemporary Caucasian woman. It means that after twenty-four thousand years, that particular group of human beings expressed by this artifact has taken no change on their facial feature. This further means that the same stabilization of facial feature can be retracted backward for another twenty-four thousand years. So here is another evidence to put the out-of-Africa theory in doubt. However, this topic has shared enough discussion in this article. Let us pay attention to something else.

Fig. 21
Courtesy of *Smithsonian Intimate Guide to Human Origins,* by *Carl Zimmer*

h. Migration, Nature, Art

In those primitive days, when the ancestors of hominids must depend on the food supply that was only provided by nature in the wild, seasonal migration must be a common event for them. The not yet consolidated island belt offered very convenient voyage route for them. Each migration caravan might not cover the entire distance from Gibraltar to the Far East Asian coast or from the north shore to the south shore of Mediterranean Sea, but covering two by one thousand kilometers or two by two thousand kilometers per year for many groups should not be a task with too much difficulty. Some caravans may even cover broader ranges than these. Migration territories overlapping by groups coming from different locals should be very frequent.

During the overlapping, both gene exchange and gene elimination naturally took place. However, the intense migration overlapping between hominids and the subsequent genetic exchange resulted in something very natural: monkeys and apes between African and Asian look outstandingly different, but comparatively, human beings have more remarkable resemblance between those found in the Mediterranean region and those found in the Far East Asian coast. With their better-developed feet for land traveling and better retained body fat for the aqua voyages, the primate species leading to the appearance of hominids had higher mobility than the monkeys and apes. Therefore, higher resemblance within hominids than within monkeys or arboreal apes is resulted. Their frequent and intense migration may also have been one reason for one common linguistic phenomenon: today, from West to East, many ethnic groups have almost the same elementary pronunciation in their language for mother, which sounds like "ma." In contrast, the equally important word for "father" in modern society sounds widely different between locations. The reason is that the need for a sound for "father" in human society appeared much later after hominids lost the need and convenience for intense migration.

Among the seasonal migrations, those moving north and south bounds should happen with more intensity, more frequency, and being more voluntarily than those moving between east and west bounds. This pattern of migration can be derived from the migration made by the nowadays migrating birds and migrating ocean animals. The major reason were basically two: (1) a more comfortable temperature and (2) animal as food source were also migrating and plants as food grew and perished seasonally.

Gradually, the dramatic seasonal scene change along the migration routes added a stronger and stronger stimulation to the visual sense of the hominids. With their pair of eyes that possessed the ability to discern spatial depth and to intercept a broad color range in the optical spectrum, with their highly developed intelligence, they began to feedback to the environment with the information they visually absorbed from the surrounding. Art expression became a strong desire. DNA mutation for the appearance of art cell in hominid brains might not have to be a necessary factor for the different artistic expression desire to emerge. It was the quantity of accumulation of stimulation and experience of numerous generations of an ethnic group that made the difference.

Because of the emerging of desire of artistic expression, which gradually elevated to ideology, the appearance of hominid feature was in turn inevitably influenced by certain of their artistic preference. This is to say that, ever since hominids showed desire of artistic expression, natural selection was no longer the only force to modify the physical traits of human beings. Cultural selection exerted force on such modification too. There have been two ways for cultural selection to work: external body part alteration and genetic selection. External body part includes body piercing, body painting, circumcision, tattoo, feet binding or something similar done in the old days, and visiting beauty salon or visiting plastic surgery clinic to copy certain trait from the others done in the modern days. Genetic selection is realized through mostly sexual interaction, of course, but also in some extend reinforced by genetic elimination through biased emotional judgment. Genetic selection, if through reproduction, has always been guided by the hope of maximizing the chance of getting the best combination of power/wealth as well as healthy offsprings. So the physique of each individual played an important component in such selection. Typically, for example, in the European area, the following trait may be important in the list for selection in the old days:

1. *Pink facial color.* A sign conceived as being healthy as well as an indication of a background of a more comfortable social condition. A pink cheek is shown more vividly with a backdrop of skin color that can reflect more chromatic components of an optic spectrum. Human beings subconsciously prefer visual expression of richer chromatic components. This is why chromatic TV is found more popular in department stores.

2. *Less-protruding jaw.* A sign conceived as less need for frequent hunting, particularly the barbarian hunting with jaws, i.e., an indication of a more elite social background. We humans all came from someone with protruding jaws (or even muzzle), wide mouth opening and quite distantly separated nostrils. Without backup of force derived by cultural development, no human can afford to have a less barbarian jaw and mouth in nature.

3. *Wider eye opening.* A sign conceived as being able to sweep a bigger visual angle without moving the head, thus more alertness in the wildness.

4. *Cold-color iris, such as blue, green, grayish.* Mother Nature seems always leaving iris in color other than dark brown for various animals that carry lighter color of body surface (skin, feather, hair). This may be to suit the need of a better camouflage. However, if a person carries a warm color iris, such as red for a white hare, or bright orange for some lemurs, in a human society, it must be offensive to the feeling of the others and will be easily excluded from the chance of genetic selection.

5. *A body with strong and healthy appearance.* Of course, this was the most important factor in the old day to every ethnic group. It directly related to material procuring, ability to rear and support the future generation, territory defense, and ruling over the others.

The existence of frequent north—and southbound migrations of hominids in the Mediterranean region of the old time is evidenced by the look of the Middle East people. Placing the physical feature of European people at one end of the physique spectrum and people from the African heartland at another end, we would find that people around the Middle East possess features that could be considered as mixture of these two extremes: their skin color is of a medium hue between whites and blacks, their hair form is more relaxing than the tightly curled hair of African people but not as straight as European people's, their nose is tall and pointed but the nose base may be quite broad too. As to their eyes, since there is no too much difference on the eye shape between European people and African people, so are the eyes among the Middle East people. Compared to the eye shape of the Far East Asians, Middle East people's eye could be easily grouped with closer kinship to Europeans and blacks than to Far East Asians. While the physical features of the Middle East people are found in the middle of a spectrum, they are also found geographically located between the whites in the north and the blacks in the south.

Nevertheless, however, it seems that there is more resemblance between Middle East people and European people than between Middle East people and people from the African heartland. That was because Sahara began to forbid the migration route deep into the African heartland one million years ago.

There is also one characteristic that makes the Middle East people uniquely stand closer to the European and African people than to the people from the Far East Asian coast. It is quite apparent that the average males of European, black, and Middle East people have thicker facial hair and denser body hair than the Far East Asians males. Interestingly, however, when comparing the females between all these people, the Mediterranean females, including blacks, do not show conspicuously more hairy than females from the Far East Asians. If hair, or fur, is fundamentally a result of resisting insect biting, this could be a hint that there was a long history in areas surrounding the Mediterranean Sea, males extensively went for hunting on open land while their female partners or harems caring their babies at homes that were still in water. Water provided a shielding action for the females against the bloodsucking insects while open field let the insect have an easier access to the males' skin. Of course, the similar situation might exist in the Far East Asia, but the violent weather condition in the Far East Asia might have reduced the intensity of insect swarming there. The higher concentration of sickle cell phenomenon among the sub-Sahara people should serve to give us a clue that bloodsucking insects did show more menacing in the Ancient Mediterranean Sea region. The same reasoning may also give us a hint to unlock the puzzle why human males usually get far more facial hair than the females. In the old time, male must hunt more extensively than females. There was a long period of time in the primate history in which jaw was their major weapon to make a kill. Naturally, bloodstain on the face was almost a body characteristic by birth to the males in the field. Blood drawn the swarming of insect; thick facial hair was thus needed for protection. As to the female, not only they made less direct hunting because of the males' contribution, but also they were closer to water more often, and bloodstain could be cleaned easily and frequently.

While the external body-part alteration for any ethnic group has never been giving any impact on how a new born may look, the genetic selection based on cultural reasoning has always been, although one generation at a time. While genetic selection was guided by hope of improving social condition, it may have been more and more compromised by ideology in nowadays society too. Such compromising may bring negative effect to human society, as it introduced chance of preserving bad gene. Cultural selection is bound to give human civilization an explosive impact someday.

i. Fire, Tool, Jaw, and Civilization

Among those hominids leading to the descendants of blacks, whites, and Asians, it is hard to say which group used fire first. It is even hard to determine that Neanderthal had not use fire ahead of our ancestors. One thing that could be certain is that those living in the more northern areas had one more reason in desiring to get access to fire than the others: to get warmed. In the old time, they needed the fire for warmth during cold time, or even during some cold nights in warm seasons. This led those northern ancestors to more frequent usage of fire. The illumination effect of fire subsequently extended the brighter duration of the day for them.

No matter where a person is located on earth, the average amount of daylight time is the same for everyone in a year, although daylight intensity may be different according to different latitudes. The bright period of the day extended by the usage of fire did come out as a big advantage to them for information exchange and civilization accumulation. This extra opportunity of information exchange and accumulation might not have anything to do to result in a bigger brain for those hominids. However, among the hominids in similar situation, there were rival groups. Those times were eras dominated by rule of jungle. If the offsprings of a certain group of hominids could not have brain big enough to store and process the increasingly accumulated information, they might be wiped out by other rival group. Compared to other animals, hominids gradually put more and more of the brain factor to intensify the rule of jungle. Fire usage certainly triggered such brain factor to become more dominant in their life.

In the section of "Hominids" (chapter 2, h), we mentioned that the talent of tool using must be accompanied with each batch of aqua hominids in the transition of becoming land hominids. However, the talent of tool using might be locally restricted if information exchange, or communication, was absent from their social life. Without advanced communication skill, the tool using talent among chimpanzees is limited to links between mother and child. With extra opportunity offered by fire to develop their communication skill, hominids spread their talent within a much wider network of social members. Fire usage therefore played an irreplaceable role in escalating civilization.

Not only has fire played a critical role in advancing hominids' civilization, but it must have also modified our facial features too. More specifically, tool using beginning in the early hominid history gradually modified our jaws. All mammals, carnivores or herbivores, before they can swallow a small morsel of food, they must tear that small piece apart from a bulky food source with their jaws, whether with the help of their forelimbs or not. In addition, carnivores must use their jaws as fatal weapons too. As such, a strong and a protruding jawbone and, subsequently, a strong and barbarian snout is imperatively important for their surviving. This was once true for the ancestors of all primates too. Indeed, one jawbone unearthed as belonging to a hominid (*Australopithecus robustus*) showed such a barbarian structure, with long and narrow mandible and big teeth that only a bone crusher needed it. Gradually, however, this became less so because they used tools to replace the jaws for the same job: food clipping, hunting, or fighting.

So weaker jawbones, followed by shortened snout, were gradually tolerated and accepted by Mother Nature for the primates that entered the hominid stage. While tool usage definitely provided chance for the hominids to weaken their jawbones, food softened by fire usage alone might not have been equally decisive in affecting the shortening of the snout. It is the elimination of using jaw as a forceful tool or weapon that had such a decisive influence. Let us present a metaphor here. In fitness room A, one group of people has exercise of constant weight lifting of ten kilograms; another group does the same thing but with only one kilogram. In fitness room B, one group of people has exercise of constant weight lifting of one hundred kilograms; another group does the same thing but with only ten kilograms. Although the multiplication factor is ten times on the weight between the two groups of people in each room, the arm thickness thus resulted in each room would be quite different. While arm thickness between the two groups in room A may not be obviously different, the arm thickness between the two groups in room B could be strikingly different. Exercising the jaws by chewing raw food or cooked food may compare to weight lifting in room A; exercising the jaws by using it as a forceful tool and weapon or by chewing raw food may compare to weight lifting in room B. So jaws that matches in the exercise pattern of room B stay stronger. A weakened jaw is definitely a negative degrading development for surviving in the wild, but hominids' civilized life had made Mother Nature tolerate it. Not only that, their cultural selection may even speed up the reduction of prominent barbarian look of their jaws too. Following the shortened snout is naturally the development of smaller teeth. Comparing between Caucasians and Neanderthals, Caucasians seem to have a more pointed chin. It is not necessary that Neanderthal showed less mass at the chin, or a shorter length at the bottom line of the mandible, it may be that the Caucasians have had their dentition receded more backward as a result of cultural selection.

j. Venus Figurines

All Venus figurines discovered in Europe were of women that were almost unexceptionally oversize, even judged with standard of today's society, which is supposedly more materially abundant than the societies of thirty thousand years ago. The majority of these figurines were found produced twenty-two thousand years or more before today. A well-done piece Venus of Willendorf was about twenty-five thousand years old; while the oldest, so far, Venus of Hohle Fels, was about forty thousand years old. They were thus produced in a warmer time period preceding the last Ice Age. About the same period of time, cave paintings were also found flourishing. With the art skill detected from the cave painting, probably it is fair to say that these overwhelmingly fleshy figures may not have been any kind of exaggeration but reality presented by the old-time artists.

A woman with the body figure depicted by these figurines could not have an active part in hunting, at least not too often at all. Indeed, if she ever appeared in the hunting field, her comrades would have to pay more attention to her safety than hunting. Similar figures are found covered a broad range of land, from France to Russia. So their meaty figures should be more of a depiction of reality other than a religious wish by people spreading in such a big area. That women did not have to hunt was made possible only if the men could bring back enough food. So combined with the fact that animals are found so dominantly occupying in the cave paintings, we may imagine that Europe in those times may be a land of plenty. Without too much effort, men could bring back more than enough.

With the body type we learn from these figurines, we may also imagine that human beings in this region relied on animal as their absolute main diet. Indeed, human beings possessing these figures may even be descendants of someone who had been completely isolated from aqua life not too long ago. (One hundred thousand or two hundred thousand years are not too long in evolution history.) More often than not, the physique of aqua animals shows more roundness, and the physique of terrestrial animals may show more pronouncing muscle tone. The physiques of these women thus strongly related human beings with a past aqua life. With so much resource naturally available, food competition between ethnic groups needed not have to be fierce. As soon as each group did not over invade other's territory, war may not happen too often. This indirectly tell us that ancestors of modern men may not have been the culprit who pushed the Neanderthal to extinct, regardless who had been more intelligent or more able. They may not have inflicted direct violence on the Neanderthals, and they may not have depleted their food source either. Some other factors may have contributed to the Neanderthals' extinction.

The last Neanderthal was found leaving the world about thirty-two thousand years ago, in a time human ancestor felt they were in a land of plenty. If we look at the skeleton of the Neanderthal, we found that their abdominal area should have been a fat storage tank. Meat eating was favorable to fat accumulation. Neanderthals were meat eater. On a land that was plenty, they did not have to struggle hard to get food. Subsequently they did not have enough physical exercise to dissipate the excessive energy they built up in their body. A warmer climate did not demand too much of their body energy to spend on cold resistance either. Gradually, heart failure became a misfortune that they could not shake off. There may be another factor contribute to their extinction. Their average height was about five feet six inches for the males, and even about five feet for the females. These are not very big frames, but their adults had a brain size of two hundred cubic centimeters more than ours. So their babies may always have unusually big heads for the birth canal of not-so-big mothers; birth complication may be a frequent accident that only modern medical technology can solve. In their time, brain factor was still an important factor in supporting the survival under the rule of jungle among *Homo sapiens*. With their brain size, it may not be necessarily justified to speculate that their intelligence was the pronounced reason for them to be defeated by human being's ancestors if conflicts must occur between them. It could even be that they were smarter than human's ancestors, but the leading edge was not enough to overcome the disadvantage brought up by a smaller body in comparison to the much bigger human's ancestor.

Part II

Cultural Evolution

(In all of the following text, the word "right" is to mean "enjoyment" or "entitlement." It is not to mean righteousness or correctness, neither to mean the direction opposite to left, unless contrast is immediately presented in text adjacent to it.)

Chapter 5

History of Tomorrow

If someday some particular race no longer shows up on earth for any reason but this race was once owning a big population, should human being's evolution be considered having taken a sharp turn in its evolution path? Why not?

One of the obvious well-known facts is that Caucasians' population shows a more and more dragging replenishing rate, even negative in Europe; in contrast to this, other races all over the world have a prolifically positive rate. A mathematical asymptote of zero is almost the only choice for the Caucasians' population. Some events may easily play a role as catalyst in the future to accelerate the approach of this asymptote. Luckily to them, by the time such an asymptote becomes a reality, they would have wisely shammed a little of their genes into the genes of other races to be piggybacked. Some term has been devised by some intellects in describing such a "free" gene riding. They call it human browning. Isn't it full of scholastic wisdom?

People in some future generation may find comment in history books similar to the following: "Thanks to the Caucasians for their abundant inventions, creations, and discoveries, in particular, their weapons, with which they had repetitively decimated each other so effectively." Some of the anthropologies of our generation hold the view that *Homo sapiens*, with the more advanced intelligence, had pushed the Neanderthals to extinct. In comparison to this view, someone in the future may have full right to write the following: "We, with unmatched intelligence, have brought up the extinction of the Caucasians." Is the famous motto familiar to everyone that "history is written by victors" (how about by those who survive)?

Caucasians have pioneered the Western culture for many centuries. This is a conclusion needs no dispute. Who will continue to carry on the Western culture after they bow out? Don't worry, the Western culture will not completely die with Caucasians; it nevertheless has some extremely juicy part, such as the medical methodology. However, can the Western culture escape from taking a submissive stand when it appears to be intolerable to some anti-West but dominating ideologies in the future? For example, can Christianity, a belief popular among Caucasians, pull itself away from the top of a list of witch hunt? Indeed, such witch hunt already happened: Christian clergymen and followers are always harassed or even arrested under various "reasons" in Communist China, and Christian belief is even life risking in some Islamic countries. In comparison, Muslims and Buddha followers are not found facing such threat in any country where Christianity is popular. Believe or not, even in America, Christianity has been under ever-increasing political pressure.

Caucasians' population in America, so far, keeps a positive birthrate. However, it is a rate far behind the rate of most of the nonwhites on this land. It is said that in 1950, about 87 percent of the American population is of Caucasians; it is also expected that in 2050 less than 50 percent of the American population will be of Caucasians. With this rate of shifting between racial populations, it is only a matter of time that Caucasians will be drowned in the gene pool of other races. How much longer will it take for them to lose another 37 percent from the entire population? It will not need another one hundred years! Coupled with the rapidly deteriorating economic production capability that we have seen in America, someday, photos from America are bound to constantly confuse the pupils worldwide in spelling the word "America."

"Teacher," students ask, "when must I use 'me' in place of f in spelling the name for the same land?"

"It is not quite the same land, children," answers the teacher. "They are two separate continents but almost look alike, except that America is more war torn. Nevertheless, America was once the richest country in the world. Now, much of her wealth is long gone."

"Is this the reason why only four or five stars are left on her flag, and all the others got lost?" asked one of them.

"That is only one of the many flags you see on this land. Instead of being rich of substance, she is now rich of flags."

As to Israel, at the time that the above questions and answers go back and forth, good luck, archeologists!

Finally, with too much hesitation, history brings human beings the chance to settle the argument between theism and atheism. Make no mistake, however, only followers of two ideology camps can qualify themselves for the settlement: Islam and Socialism. And only one single issue needs to be settled: whether God is in heaven or on Earth. To Islamic followers, "there is no god but God"; God is in heaven. To Socialists, "there is no god but me—God"; God is right down on Earth. To Islamic followers, diligence must be shown to Quran so that someday Allah comes to admire us all. To Socialists, diligence must be shown to Marxism so that someday no one is allowed to show any disagreement as they "liberate" the entire mankind.

The West is said on the way to meet her death; the United States of America is said waiting in line for her bankruptcy. These have been open secrets to the world, but only two groups of people are found not being aware of them. The first group of people are those who have had their eyes blindfolded by all kinds of "rights" that they demand their governments to deliver. The second group of people is those who know they will lucratively profit from the collapsing of the West. Of course, then, these open secrets are just pretended not being seen in the forum of the second group. Inside this group, fellows share with each other the true nature of all the good news about the West's collapsing; but why wake up the suckers? In fact, do invest in it when possible! They may even create various massage parlors for the first group of people. Inside the massage parlors, they encourage the first group of people who crawl on the massage table comfortably: "There should be more rights waiting for your enjoyment, baby. If you whine hard enough, the government will let you have them . . . Oh no, don't have to pull your thumbs out yet. Suck a little longer . . . Life is hard outside, so prepare yourself better. Pursuing happiness is your inalienable right, it needs a clear mind, well-fit body, strong will, and noble spirit . . . Let this humble parlor prepare you well for all of these."

The open secret that the West is dying includes two parts: the open part and the secret part. The open part is that she is dying, seemingly incurable. The secret part is that, at her own expense, she has endlessly created paradises to entertain all culprits who can only lead her to meet her grave. The number one culprit she needs dearly is cheap labor. Without cheap labor, she will die of hunger; with cheap labor, however, she will die of blood depletion.

Chapter 6

Labor Monopolization:
A Devil Twin of Capitalism

Workers on All Land Unite!
—Karl Marx

a. Two Production Partners

The process of production activity needs two partners and two partners only: capital and labor. You can call them twins. Cash, raw materials, equipments, and infrastructures are all capitals although one may be more liquidating than the other. They cannot convert themselves into anything desirable until the alchemist—labor force—touches them. If either partner, or either brother of the twins, stops its continuous supply, the production activity must also stop. However, labor is always acting as a cost to capital, while capital with careful management is a beneficial element toward which labor is contributed itself for. So it can be imagined that if labor is not cheap but expensive, capital will be clipped in some extent; in turn, labor must eventually enjoy less of what capital can provide. The simple mathematics is that the more expensive the labor force becomes, the less the capital can be accumulated. When the overall increment of price of labor force surpasses that of the growth of capital, a peril begins. The direct interpretation of such a peril is that capitalism gets choked. When some threshold strength of the choking is reached and nobody is willing to let loose, endurance of the society is gone, and death descends upon capitalism. To end the capitalism, a violent revolution is not even needed, but a choking caused by slow price gouging from the labor force can do the job. Relax, Marx! While Marx was exactly wrong in advocating a concept of "dictatorship of proletariat," he showed high accuracy in predicting how capitalism may get destroyed: "What the bourgeoisie, therefore, produces, above all, are its own grave-diggers." Either through bloodshedding or price gouging, grave digging is under way.

Labor force is like any merchandise in a market. If it is in a truly free market, competition between labor providers will never enable it to surpass the capital accumulation. It will, however, if competition is removed, and shoppers of labor force have no choice. Monopoly on labor will do just that: leaving the shoppers of labor forces no choice. That monopoly of capital will destroy the social harmony in capitalist society has long been recognized. To avoid the danger caused by such monopoly, laws after laws have been enacted by capitalist governments. While the governments work hard adjusting the balance on the capital end, the other end of the same balance has escaped the attention: labor force is getting more and more concentrated into someone's hand, and monopoly of labor force gradually shows up.

Labor force in capitalist society began its tendency to be monopolized as early as at the time Karl Marx called on that "workers of all land unite!" The obvious result of the labor monopoly, besides all its other social impacts, is that labor becomes more and more outstandingly expensive, and subsequently, capital accumulation in the society gets more and more amputated. This grave effect has been no more apparent in nowadays Western capitalist societies. Hasn't the collapse of the American automobile industry offered a good evidence to illustrate the problem? This is only one of the bigger problems that follow many other less obvious problems that have occurred. For example, how many Americans would find what they wear is "made in USA," or how many American families have TV set that is still "made in USA"? Difficulty of capital accumulation in America has eradicated the production lines one by one from this land.

When domestic labor has been monopolized, there are only few choices left for the capital holders to overcome the difficulty: (1) to get the production going on at the mercy of the labor monopoly groups any way, betting that the monopoly groups also want to keep the hen alive for eggs continuously coming; (2) to bring in cheap labor from foreign counties; (3) to have the production activity operated abroad, in other words, to send the industries oversea, where cheap labor can be recruited with much more easiness. Way 1 is obviously not on a way to favor capital accumulation because any possible capital accumulation is seen as a result of having had the selling price of labor unreasonably lowered. Monopoly of labor is exactly devised for the prevention of such thing from happening. Way 2 has resulted in the rapid expansion of Islamic territory in Europe and a completely irreparable floodgate in the American southern border. Way 3 results in accelerating accumulation of capital at an astonishing speed, but in other countries, such as South Korea, India, especially Communist China, where the most abundant supply of cheap labor in the world can be found. All three ways must work indifferently: plunging the originally developed capitalist countries deeper to their graves at higher and higher speed.

Now, let us see what way 2 has brought up.

b. "Cheap Labor," Irreplaceably High Cost

If it is not because of the three major events of population disintegration (Russian Bolshevik movement and the two world wars, we will discuss them in more details in later chapters), Europeans may have been able to replenish their own cheap labor for a longer duration. Their capitalist economy, as well as the surviving of the Caucasian race, may not have to feel their final gasp so intensely as what they feel now, at least not that quickly yet. During the long period between sixteenth century and twentieth century, the Caucasians in the American continent needed to import cheap labor from Africa and Asia; the Caucasians in Europe at the same time did not feel the same need. Indeed, Europe even had excessive cheap labor to export to the New World during the same time. However, after WWII, the Caucasians in the Western Europe began to feel the hunger of cheap labor. If the Communist Soviet Bloc had not been so persisting on its ideology but allowing a free immigration policy, the Western Europe's hunger may be subdued in some extent because cheap labor may keep coming from the Eastern Europe. Now, the Western Europe had to satisfy her craving from another source; and people hungry of work in Islamic world were waving hands almost immediately next door. Supply and demand on labor force matched quickly.

However, are the cheap labors in the Europeans' concept really that cheap after they recruited them from abroad? Do they need to offer something else beyond monetary reward to make them happy? How long will they stay cheap? Some of the European intelligent elites soon found something that is almost costless in money terms. How about the permanent residence and citizenship? "Um, good trade," thought they. In order to maintain the influx of cheap labor, subsequent modification of immigration law becomes necessary. In little than half a century, many habitats of the original European indigenous become territories of different ethnic groups'; the original European language, a big body of European laws, habits, traditions, and sometimes even some European individuals are not found respectful but repellent. In some European countries, although not popular yet, it has happened that more than half of the newborn babies are of Islamic descendants. With the population landscape to be replaced in such a speed, the power structure in the European society must be soon anew.

At the beginning, European people held some hope on assimilation. More often than not, however, hope of assimilation walks in an unexpectedly reversed direction. The anticipation to see more school girls wearing pendent of cross is replaced by challenge why a hijab cannot enjoy the equal right like the cross pendant. Outcry of autonomy government prevails at habitats where residents of European ancestral background are found having become minority. In Denmark, it is said that an Islamic family with a newborn girl would receive an assignment from someone unknown, telling the family that this girl must in the future marry with nobody but someone from some Islamic country—*or else*! Is it a rumor, or is it factual? Practice on how Islamic world handles religion conversion should give us a hint. Conversion from Christianity to any other religion is not found restricted anywhere in the world, but conversion from Islamic to any other religion is basically forbidden in Muslim dominant country. Cost can be up to death for such conversion. Apostate and infidel are equally offensive to this belief. What happens if someday Islamic population becomes dominating in some European country? Why should the same practice on religion conversion be given up only because it is in Europe? If they want to influence on religion conversion, one can bet they will want no less influence on marriage. A successful influence on marriage must greatly benefit the expansion of their habitats.

Considering what the Europeans may have to ultimately lose, who can say that Europeans are not currently mortgaging what their ancestors passed to them in exchange for cheap labor? The commodity of land cannot be reproduced in the market, while the commodity of labor can be limitlessly replenished; and more interestingly, each of its reproduction is accompanied with joy. If a reproducible commodity is not to be acquired unless an irreproducible commodity must be paid in addition to money, how good and fair is such a purchase? Why is it that mathematics cannot make sense on this land? The answer may rest on that Arabic nations have ancestors who were outstandingly intelligent in mathematics. When Egyptians built their pyramids with nearly impeccable accuracy, Europeans were not found building anything significant yet. If the Europeans eventually lose their dominance in their own habitats, their future generations must groan with grief "Why had our forefathers paid somebody to do all these to us?" Too bad, the effect from the three population disintegrating events has set their destined course; too bad, when their fore parents were so occupied by the good feeling of comfort, the welfare of the children was left out for calculation. Miracle is needed to reverse the course, and the miracle must include the restoration of strong will and the removal of labor force monopoly.

Similarly, cheap labor in America is also expanding its territories without bound. Over here, such expansion is not signified by religion (but is on the way), but language. Its effect on degrading the cohesion of a nation is no less than religion; in some situation, it is even more intensive. However, those who outcry "wake up, Americans, you'll lose your country if you don't stop them [the illegal intruders]" must know that their effort will only end up defeated. It doesn't matter how loudly they cry; if they cannot overcome the monopoly of labor force in America, they cannot overcome the illegal immigrant problem. Labor monopoly, political groups with special ambition, and some interest groups already entwined together as a formidable political conglomeration, aiming at usurping power of the nation. No better political force can match more perfectly to their agenda than the godsend cheap labor troop from South America; no better political force can be so effortlessly recruited and bought with the treasure from others.

Similar conglomerations existed all over the world; Socialism movement continues. For the movement to be pushed forward, the illiterate cheap labor force is the best force to be utilized. Raising the same flag of pursuing "fairness" for people, but dressing themselves differently, members in these political conglomerations are pursuing the same agenda as what the Socialist forerunners were pursuing in the first half of the last century: monopolizing absolute power. Those with a guerrilla beret cap openly declare themselves as Socialists, as this is a fashionable term in the basically illiterate and economically backward areas. In these areas, gun and blood are what they believe. Those with suit and tie name themselves as diligent workers for the grass root, as they know the term "socialism" has no attraction in areas where higher education is more popular. In these areas, the ballot box is their tool. So in the so-called democratic societies, all what the suit-and-ties need to do is to keep the door open for the cheap labor to pour in and then to politically "baptize" them with proper political terms. That is the reason why a sturdy fence along the American southern border appears so difficult to erect. With America's financial ability few years ago, even Mount Everest can be put there, not to mention a fence. It is the problem of will, not the problem of money. If the labor monopoly group, the politicians of special ambition, and some interest groups cannot be untwined from this mighty political conglomeration, protecting the entirety of this country is an uphill battling on a very steep slope.

c. Grave Digging

The democratic society has laws to break the monopoly on capital so that competition between capitalists is as fierce as one can see. On the contrary, however, it has law to guarantee the monopoly of labor. For example, the right to freedom of association is recognized as human right, which is a political freedom and a civil liberty guaranteed by constitutions in various countries, and is therefore almost untouchable. So we can see that law has unfairly adjusted the balance on the production parity. This means that capitalist society has willfully made laws for self-choking. So production is put between a pair of blades of the mathematical scissors: a product must be sold at the lowest price while it has been made at the highest cost. As the blade of price limitlessly pressed down by the market, the blade of labor cost is uncontrollably risen by the monopoly of labor force. These two blades must meet but at one point where production is guillotined. The failure of GM is no way the first victim of this pair of scissors.

In the early 1950s, America controlled 40 percent of the merchant fleets in the world, where have they been now? GM will not be the last victim either. The problem is that as the list of victims piling up, America is brought to her knee faster and faster. Of course, if monopoly of capital succeeds in its agenda, the pair of mathematical scissors flips in another way: a product is sold at the highest possible price while it has been made at the lowest possible cost. Then, labor force gets guillotined. Believe or not, this is exactly what happens in all the Socialist countries; because the leading (ruling) but the only party monopolizes everything there, capital monopoly is only one of the item. In the twentieth centuries, pioneered by Theodore Roosevelt, Western governments have kept their hawk eyes on the scissors of cost and price. However, so far, only the capital end has been watched; the other end of the balance is all time neglected.

Very plain mathematics has taught more and more political candidates that showing allegiance to the sources of labor, having been monopolized or not yet, can earn them more ballots. The classic view about the Western capitalist governments, typically found in Marxist's writings, is that they all are agents or puppets of the powerful capitalists. The time to revise such classic view has long been overdue. As the capitalists must more and more frequently kneel at the labor monopoly groups, siding with the capitalists is no longer a smart political calculation for those looking for government jobs. As a matter of fact, siding with the sources of labor, these applicants of political jobs can even count on the business owners to send them into the offices too. Particularly for those business owners who feel unable to move their businesses oversea but staying home being the only choice, they are so desperately to hope to see cheaper labor to set foot at their employment offices. So strangely, but naturally enough, two groups of opposite interest in economy put themselves in the same alliance, in need of the same government officers to realize their goals. It is a dangerous combination! Government is thus abducted without being so officially declared; one single formidable political bulldozer is thus assembled.

The end result under the mowing of this bulldozer is that the traditional industries are uprooted from the land one by one. Those businesses that can escape to oversea will leave, as mentioned in way 3, section a, chapter 6; those cannot escape will sooner or later die. The business owners who form alliance with the monopolized labor through the political agents eventually cannot protect themselves; their count of ballots is always at the minor and subsequently the losing end. Way 3 also adds another graving effect to the nation: to bring back products that are produced somewhere else, then, not only sucking away the domestic financial source, but also further pressing the similar domestic industry to meet death. To compensate the financial vacuum created by the disappearance of the traditional industries, nothing is found being more satisfactory than mortgaging out national heritage. Mortgaging national heritage makes no one angry. National heritage is something passed down from great-grandparents; it costs nothing directly from anyone's pocket.

The national heritage includes, but not limited to, country sovereignty, national language, and national real estate ownership—citizenship.

Here comes good examples how mortgaging national heritage is found emotionally immune in America: Waves after waves of demonstration by illegal immigrants holding foreign flags are seen marching on the American land, no equivalent counter demonstration is ever seen; bills after bills to make Spanish a semiofficial language in governmental offices are generated at various level, no single significant bill to counter such diligent effort has even been found. However, when tax money is seen improperly spent in various stimulus packages and bail out programs, Tea Parties show up to respond in a short time. Why? Stimulus packages and bailout program are seen more as hands directly reaching into someone's pocket. National heritage is far more valuable and fundamental than tax money to guarantee all these citizens' future and should have drawn more serious attention way earlier. But why were the responses against the mortgaging on national heritage have been found so retarded compared to the responses to the pressure from tax money? Mathematics once again fails to make sense in the Western culture. It is so amazing! Who can say that the appearance of Archimedes of Syracuse, Isaac Newton, Wright brothers in this culture is not merely a coincidence? Obviously the formidable political conglomeration has chosen the right spot to begin their biting: the area that is more profitable but less emotionally agitating. As the similar mortgage biting in both Europe and America is in progress, citizens' traditional habitats are seen continuously shrinking, population dominance of traditional religion is seen continuously lowering, and the utmost primary language is seen being incessantly sidelined from all offices. All these must lead to nothing else but the vanishing of a nation!

Sure enough, some high-tech business still stays at home; *only* the traditional industries have either died or slipped away. Believe or not, if equivalently well-educated labor requiring substantially lower pay can be easily found overseas, high-tech business will slip away at no time too. Quite a few years back, one California utility company needed drafting technicians. The price it must pay for a domestic one was $25/hour, with the following conditions: thirty-minute break out of an eight-hour shift, two weeks of paid vacation in a year, plus pension, medical insurance, workmanship compensation, and FICA from the government. For the drafting person of the same quality found in Thailand, all this company had to pay was $40 a day, but absolutely nothing else, with the condition that certain things must be completed by a certain day. How should a business owner make choice in between?

Alliance as what they have formed, the business owners at home and the labor monopoly groups nevertheless act in a conflicting manner. As soon as the business owners lure the appearance of cheap labor, the monopoly groups will try every effort to circle them in their monopoly barns. Then such a cycle begins with increasing magnitude: more cheap labors are needed to satisfy the craving of the business owner, but as soon as they appear, they will be circled into the monopoly groups' barns; then more cheap labor must be sucked in again for the business owner. Along with each current of the cycle is more and more of the national heritage to be scrapped up and mortgaged away. So far, the mortgaging still goes on well. Europe and North America, Australia and New Zealand next in line, have concentrated the most valuable real estates in the world. So many people all over the world are so willing to risk anything, even their lives, in order to get access to these lands. But how much longer can real estate support the mortgaging?

Now, those who cry for the defense of national heritage no longer have to wonder why they never have enough headcount on their side and never enough monetary resources to support their agenda. They must have also figured out by now why so many media networks choose to be on the side opposite to theirs. The simple political mathematics has placed many media networks in a position of special interest. Their domestic sector must choose alliance. Not only making alliance with the political conglomeration that is mortgaging the national heritage provides them with strong support, but the increasing population of the cheap labors also provides them with an ever-expanding audience market for their advertising revenue. National heritage? Why care! The loss of national heritage is everyone's suffering, but the "success of my business is mine to keep."

Not until the grip of labor monopoly is pried loose can the foreign cheap labor be stopped from pouring in; not until the grip of labor monopoly is pried loose can the production line be stopped from slipping away overseas; not until the grip of labor monopoly is pried loose can the national heritage be curbed from being incessantly mortgaged away. Make no mistake however; to untie the grip does not mean to completely remove the association of labors. Labor unions must be kept. Otherwise, it is another way to urge the restoration of capital monopoly. What should be pursued is to stop the absolute dominance of a single labor organization over one single company. Comparing with that law is there to limit the size of capital controlling capacity of some business owners, law must also be set up to limit the size of the labor organizations, in the national level, local level, and in the level of each production line, each professional occupation, and company. Otherwise, carrying the case to extreme, if a government must be reined by forces from Republicans, Democratic, Independents, Conservatives, Liberals, Lefties, Right-wingers, Christians, Socialists, but all national human resources can be commanded by one centralized will that is independent of government, wouldn't it be time that this government is to be replaced? Workers' willingness to work as nonunion member must be fully protected by law. If Constitution protects one's right to join an assembly, why shall not it protect one's right not to join any assembly with even more vigor? Competition between capitalists is vital to the society; no less vital is the competition between labor forces. No one should be given the power with which he can make the society kneel at him or even paralyze the society. Square deal advocated by T. Roosevelt is not enough; a true balance on the production parity must be set up to calm the society.

It is time for those who believe in the monopoly of labor to think which of the following ways will bring them more profit out of the killing grip: further escalating the grip strength to eventually get 100 percent of nothing from a dead body, or loosening the grip so that some percentage of something keeps flowing into the life lines. Allowing cases to be carried to extreme, monopoly of labor is far more dangerous to mankind than monopoly of capital. The reason is plain: Monopoly on capital is a monopoly of a social tool to motivate people, and the tool will do nothing if left alone; but monopoly on labor is a monopoly directly on people themselves, who is a force capable of converting a virgin land either into a paradise or into a hell. Monopoly on capital will end up with only few bosses for people at the social bottom to choose from; monopoly of labor will eventually end up with only one boss for people to submit to, but with no choice. A history of a century shown by all Socialist countries has displayed this outcome of a single boss; do people need to pay more blood to continue to prove? If you are the believer of monopoly on labor, look through your own eyes in the mirror and ask yourself honestly: Do you truly believe that you will be the lucky one who can eventually monopolize everything like all those but handful Communist gods? Or will you most likely end up as being monopolized, being finally hurled to the altar of "workers of all land unite" to meet the need of someone who has monopolized everything? Possibly it is even more horrendous for you to learn: It is not you but your children or any other loved ones who must lie on the altar with absolute tameness. As to you, you may have been wasted in the course of setting up this altar.

It is not absolutely fair, however, to say that monopoly of labor is the sole culprit in thinning the appearance of domestic cheap labor. Capitalism has another big hand to snip away her own source of cheap labor. That is the abuse of welfare system. The welfare system is devised with profound political wisdom in mind. However, instead of helping some misfortunate one to get back to his feet, this system has gradually been kneaded into a loaf of work-immune bread for some "unlucky" ones. Under this system, so many unlucky ones are so willingly to stay unlucky forever. However, do not complain about their willingness. This is the nature for every animal, including human beings. If food is constantly available next to someone's mouth, it is only stupid to spend energy to chase after something further away and harder to get. If the unlucky ones do not have to get up to look for food, how can anyone else drag them into the labor market? If it is said that the monopoly of labor damages the labor market, at least the labor forces still show up in the market when the price is "right." The damage made by the abuse of welfare system, however, is its potential of complete removal of the labor force from the market. The damage done by such abuse does not stop here, but it also converts the welfare system into a formidable evaporator of social wealth.

Cheap labor by itself is not a culprit to drag the West to the grave. It becomes the culprit through the mechanism of monopoly. Whether labor is cheap or not is on a relative scale. It can be equally called expensive labor too. That is why the term "cheap labor" or "labor force" are interchangeably used in preceding paragraphs.

Capitalism will die if its life line of capital accumulation is clipped; its life line must be clipped if cheap labor is made disappeared. To have all cheap labor hide in all "appropriate" niches, nothing else is needed but one slogan: Fairness.

Chapter 7

Fatality of Fairness

a. Concept of Fairness

Human beings are no gods; only God is eternal. With a human's limited ability, his/her capability of showing love on someone or something is only because his/her attention to other people or other things is compensated with less love, or deficit of love, or even hatred. By the same token, fairness to be made possible for some group of people is only through the compensation of unfairness that is forced onto other group of people. Fairness has been proven the most effective fuse to trigger the explosion of all kind of cinder boxes in changing routes of human history.

Everybody has the concept of fairness, but how fair is fair? Between sheep and wolf, a sheep feels unfair to be eaten by a wolf, but a wolf feels unfair not to be allowed to eat the sheep. To someone, racism against him is unfair, but should be naturally accepted by anyone else if it is in his behalf. To a Communist, the abundant income seen taking home by some capitalists should be revealed as unfair to workers. However, requesting the revelation of the more abundant wealth this Communist can spend after he gets to the throne of power is not only seen as unfair but even hostile.

The basic concept of fairness is the same enjoyment for everyone under the same condition. Unfortunately, the "same condition" can never be genuinely realized between every living being; therefore, unfairness is a norm between every living being on earth. The pursuing of fairness between the living beings must lead to disturbance, or struggle, from mild to violent. No matter how they struggle, however, at the absence of genuinely the same conditions, unfairness must persistently show up, only with a different form. Nothing can guarantee that the new unfairness will be less vicious than the old one. As such, among human beings, with their subtlety in detecting difference, constant fighting for fairness between them is induced as a norm in their entire history. Besides the instincts that are required of all animals for surviving, human beings are burdened with one more yoke: ideology, which is unshakable but unfortunately heavy to them. Before any physical fighting for fairness is brought to reality, ideology is already there to jot down the list of unfairness. As it prepares the list, it brews hatred too. No other medium other than unfairness can be found more effectively to start hatred between people, whom are grouped by different reasons, such as ethnic clans, interest alliances, classes, races, religions, countries.

Fairness governed by ideology twists people's understanding. With this twisting, one with an income of $2.00 would not feel comfortable to see his neighbor to enjoy an income of $10.00. Instead, he feels a lot easier with an income of $0.50 when he sees his neighbor with an income of no more than $0.50. The same feeling, however, will make his reasoning to be oriented in another way when social responsibility is involved: "Why should I be obliged with more responsibility than the other one?" Intentionally or unintentionally, how the others have contributed toward the society to get to where they are is not a concern by a "fairness" pursuer. In other words, fairness pursuers are quite often unfair themselves to the others because they merely focus on the same result of enjoyment but disregard the difference on conditions and effort of contributions.

Take America as an example. Pursuit of evenness in enjoyment has made her devolved into such a society in which people are allowed, or even encouraged to transfer the cost of personal enjoyment into social liability in all possible ways. At the same time, however, public responsibility is placed as distantly as possible by a big number of individuals. This is exactly the society that the fairness pursuers come after. Therefore, for example, fatherless children brought to the world by irresponsible sexual behavior flood in this country on the one hand, while military recruiters for the nation are chased away from campuses by students on the other hand. With this kind of attitude, can American people say that their soul and mind are still there for a nation to exist with good health? For evenness on enjoyment disregarding evenness on social contribution makes those who honestly devote to the society feel penalized. "Why," for example, thought a loyal father of four, "should I pay tax to support a bastard from a couple whose job is only to enjoy irresponsible sex and thus to bring in more bastards for me to support?"

It is said that one of the major reasons for America to be strong is her encouragement on competition. Competition must produce stress and anguish upon those who feel less prepared for success. In order to make the less successful ones feel happy, fairness pursuers gradually introduced a social system that either lowers the intensity of competition or rewards the less successful ones somewhat.

People can have two methods in rewarding a basketball game competition: (1) let both teams from two different backgrounds start with zero points for the match, while all players in one team is obviously at least six inches shorter than those in the other team, or (2) to grant the team of shorter players twenty points in advance before the game starts. Which method is fairer? To make it more difficult to settle is that if method 2 is habitually practiced, but someday the twenty points are no longer granted before a game starts, an accusation of unfairness must be heard. In case that the two teams have obvious racial difference, the "unfairness" may be immediately escalating to racism.

From the point of view of competition, practice 1 should match more closely to the mode of competition that happens in nature, while practice 2 has strong human intervention. Do we see signs "Do not feed animals" in many recreation areas in the wild? The purpose of these signs is not to harm the animals' natural ability in their competitive environment, as humans' "kind" intervention must jeopardize their natural ability. But when coming back to the human society, it seems that human beings always want to exert their wish to remold the exits of Mother Nature's selection sieve.

In the recent society, in order to reach fairness, practice of mode 2 is getting increasingly popular in more categories. An obvious effect of this "kindness" intervention in America is the compromise of efficiency in education. Standard of quality check is gradually removed from schools only because people from some ethnic background could not feel the same willingness of catching up. However, Mother Nature never lowers her standard in her quality check. Virus would not reduce its deadliness because the doctor is from a special race; enemy's missile will not increase its vulnerability because the oncoming antimissile has been designed by an engineer from a special race; the basketball hoop will not lower itself by six inches because a shorter player is approaching. More drastically, no any country in the international world will lower their competition threshold against America only because Americans subdue their competition standard at home. The compromising of the American education has barred American students from appearing in the top few of the lists in almost all kind of student competition in the world, mathematics in particular. It is said that China has more good students than America even has students. With this kind of competition ability, wishing to stay in the leadership in all fronts in the world is really an American dream, and dream only, for the United States of America. Sooner or later, her promising for her own willful losers must compromise this country as a willful loser in the world. Indeed, strong signs of this have shown. Do people hear that the secretary of state has been so much quieter recently when the subject of human right concerning China is mentioned? The fairness pursuers try everything to eliminate the competitive ability within Americans, while leave them inevitably facing the fierce competition in the world. Whose agenda are these fairness pursuers helping to realize?

—

b. Modern Social Effect of Fairness

In history, capitalism and Socialism see each other as enemies of no possibility of coexisting and must vow to have the other camp extirpated, through peaceful means or violent means. After some long bloody struggling, now they finally find some behavior in common: Each camp planted in some elements from the other camp that was seen evil before. So the Socialist countries, typified by Communist China, inject into their economy with the golden marketing principle of supply and demand while the capitalistic countries, typified by America and almost all Western European countries, flood into their own homes with the tragic principle of plantation: "fair" redistribution of wealth. Now, the previous vampires are seen with more rouge on their cheeks, but the formerly vibrant ladies are seen getting paler and paler and stooping more and more each day. The blood that one abundantly receives is exactly the blood that another one has been profusely losing. There is no robbing, not even one gun battle, for all these to happen. All these have happened only because some previously "fair" countries allow more "unfairness" to flow in their bodies; while those countries being complained as unfair all the time are getting more and more punched down by nails of fairness.

Ever since capitalist dominates the West, fairness has been a topmost issue for the Socialists, and later became the most beloved and invincible flag for them. Without exception, though, after they get the power, the absolute but barren fairness are only seen between the members in the very bottom of the society. Between the overwhelmingly spreading bottom tier of the society and those few powerful Socialist elites, inquiring of fairness is a crime, a grave crime. No single historical evidence has ever been found to prove the existence of the opposite. Willfully or not, however, what the various groups of fairness pursuers in the capitalist countries try to bring in is exactly this type of society. Plenty examples will be found in the discussion of various upcoming paragraphs. Only that nowadays violent revolution for the realization of fairness seems not that inevitable in the more civilized West countries. The reason is that the suicidal nature of democracy is able to be fully exploited. To make this exploitation fully possible is the cold fact of monopoly of labor, as explored in the previous chapter.

Violent or not, the ultimate social product that the fairness pursuers create is another form of unfairness. For example, in America, how fair for them to allow the illegal immigrants to have broken all kinds of law to gain legal residence in America while legal ones must line up patiently in various embassies in other countries? Currently, we even hear a term called "illegal immigrant's right" created by those fairness pursuers. Taking away the word "immigrant's," we only read "illegal right" out of this term. Alas! Fair? Nothing can be seen more plain

---

than what these fairness pursers have led to: law trampling is the way to gain "right" for somebody. When law can be applied in their behalf, they apply law; when law cannot be applied in their behalf, they toss away the law but apply fairness. This is the entire essence of all their tricks.

"Right", even acknowledged as being illegal, is now bolted down by the fairness pursuers on the land called the USA, forcing the acceptance from the people of this land, a land that has been proud of the traditional principle that no one is above the law, "the land of the free, and the home of the brave." How should one believe that this is once the land from which numerous names like these sprang up: George Washington, Benjamin Franklin, Thomas Jefferson, Abraham Lincoln, John Kennedy, Martin Luther King, and Ronald Reagan? How should one believe that this is the land whose people raised their heads even higher at gun points nearly 250 years ago? How should one believe that this is the land that nurtured a people who were willing to have their chest ripped open for the liberty of the others 150 years ago? Now, the descendants of the same heroes have devolved into political beings who must prepare themselves to be submissive to right that is openly termed as illegal. When a right is openly declared as illegal and is to be exercised over a man, but this man must paralyze himself to accept such exercise, which of the following classes will this man belong to: slave, serf, prison, lord, knight, master, and freeman?

Common sense in politics has thus been seen distorted to be inexplicable by the fairness pursuers. As right created for illegal immigrants can be seen to be made lawful, should one be surprised that illegal border intruders are called patriots? Naturally, common sense in law must make it indisputable that anyone who shows resistance to patriotic personnel can only be treated with what is for treason, a concept opposite to patriotism. With this reversing of political concepts, should people feel puzzled to learn that their border patrols be put in jail for ten or more years after they show sublime loyalty in defending the sovereignty of the country? Should people feel puzzled to learn that their policemen in many cities are threatened to be fired or even arrested if they dare to check the immigrant status of any suspect?

With a court system that is seen more frequently satisfying the issues of excessive democracy, with a Congress that shows dominantly favoring issues on behalf of monopoly of labors, with a president through whom Socialist agendas hold high hope, and with a chorus of media networks whose melody is always found composed by ideas from "fairness" pursuers, America has been seen completing her frame work for Socialist reformation. Overall, each piece of this frame work is fused together by "fairness" pursuers. What Americans should celebrate is that this drastic reformation has been processing without blood spilling, at least not yet. Following the gradual affirmation of this frame work, it is absolutely possible that what happens in Venezuela will repeat in

America, provided that the future world still allows America a chance to stay in one single piece.

It may be unbelievable that Socialists love democracy. Truly, they do; but they love it until they reach a sharp line beyond which they dominate social power. Once they usurp the social power, any democracy lover is their enemy. So, before the line is reached, they push for the agenda of democracy in all aspect lest it will not be excessive. Indeed, the more excessive the better. Why not? Democracy allows them the most extensive field for gathering social power at the lowest cost, and when they are caught red-handed, democracy offers them the best protection. A punishment against them is often cited as antidemocracy. Their various goals, which can eventually funneled to monopoly of power, the governmental power, are thus served no better by democracy. So the Socialists usually are the ones who spell the most uproar if democracy is detected with any deficiency. What they feel intolerable is the democracy showing up at where they have dominated. During the Chinese Civil War, Chinese Communist Party had official articles showed more elevated enthusiasm and more virtual admiration toward democracy than the then American leaders. In comparison, her proclaiming toward democracy dwarfed the then Chinese government in any measure at that time. One flag opens the way for Saddam Hussein to his tyranny career is the so-called Arab Socialism. Socialism has long been advocated being the best way to gain fairness to everyone. Everyone, is it really?

So far, in America, what people can see is that, slowly fused together by fairness pursuers, a social framework goes hand in hand with all of the followings: the new racism, a reversed slavery practice through an abusive welfare system, the human rights that have never been strictly defined by law but always superior to law, the incessantly replenishing influx of cheap labor from sources aiming at destroying the nation, and the limitless mortgaging of the national heritage. And then, all these come back to reinforce the frame work to be even more Socialist. And so far, as all these happen, all what people can hear is some feeble complaint: "Are these fair to us?" A message to the complainers: Who said that the fairness pursuers ever try to be fair to you?

c. Ultimate Fairness, Ultimate Monopolization

Being fair to some but unfair to some others is the typical contradicting attitudes from the fairness pursuers. The people toward whom they display unfairness are not minority in the populace by any measure; neither those people who receive their fairness treatment are of majority most of the time.

Many fairness pursers must declare capitalist society being sinful, criminal, and condemnable because they say they detect the society's allowing monopoly on capital. It is unfair, according to their propaganda, for so few people to control so much of the social wealth. At exactly the same time, however, the exact fairness pursuers advocate all agendas that must lead to nothing else but the monopoly of governmental power. Some of them do it more openly and radically, like what Lenin and Mao Tse-Tung did; some of them do it more implicitly and progressively, like what many have done in Western Europe and are doing in South America. They never explain why monopoly on governmental power will become fairer than monopoly on only capital. They do, however, tell people that monopoly of governmental power in their hands is the only way to remove the capital monopoly that people have seen; they will thus make the society fair to everyone. History has made nothing more apparent than that a group of people who can monopolize governmental power must monopolize everything; capital is only one item in their monopoly list. No one can even deduce that a group that can monopolize governmental power will intend to monopolize nothing. So "all powers to the Soviet," declared openly the radical Vladimir Lenin during the Russian Communist revolution in 1917. So in South America, some less radical fairness pursuers nationalize the media networks, nationalize the banks, the petroleum industry, the communication industry, and one step at a time. Gradually, after the less radical ones feel they have brought enough fairness to people, or have concentrated enough fairness in their hands, they move to reach the same goal as those more radical ones: to seeking presidency of unlimited term. With a president that has no term limit plus his, but nobody else's, sole governess on everything that has been "nationalized," do people need a more vivid illustration to recognize what monopoly is about, and what has been monopolized?

What makes it amazing is that those fairness pursuers who show such hypocritical advocacy in antimonopoly usually possess better education. Education must have equipped them with enough reasoning power to deduce that monopoly on government power is a much bigger monopoly than the monopoly on only capital. To an educated person, a criticism about his loss of sense in logic is unbearable. Therefore, in their list of pursuing, the fairness pursuers must keep in their mind something else that is far more worth pursuing than "fairness"; and so worth that the exposure of their loss of sense on logic becomes trivial. What is so worthwhile? The answer is exactly the same thing that they ask other people to remove: monopoly, but in their mind is a far more upscale monopoly, a monopoly of everything. After they monopolize the governmental power, who dares to show disagreement on their logic? Has anyone found a bigger "fairness" pursuer in history than Mao Tse-Tung claimed he was for the people? By the same token, has anyone found any government chief who has been able to monopolize more than Mao did in every sector of a nation?

What makes it difficult to comprehend is that so many people who make themselves uncompromising to monopoly of capital show unlimited zeal to follow those who advocate the monopoly of governmental power. Indeed, among these people, those with higher education even drumbeat louder than those with lower education. In America, one can find overwhelmingly more number of intellects, in education institutions and media network, siding with ideas that inevitably serve the course of monopoly of governmental power. This phenomenon is brought to reality by the following reasons that always interlace together:

1. In any society, it is always that fewer people control and enjoy more wealth than the bigger populace. Only liars say they can change it. Even if people are put together to form a community with absolutely the same amount of wealth between them to begin with, the next day they will find that equality of wealth ownership between them disappears. This inequality is destined by human instinct. However, being a liar on this matter and lying with a silver tongue, which is what many people investing heavily in education intend to acquire, can get someone a shortcut to power.

2. With the freedom of news, unevenness of wealth distribution in capitalist society is maximally exposed.

3. Commonly, many people feel uncomfortable with an income of, say, $2.00 while seeing another one with an income of $10.00 or even$100.00. The same people will feel much better with an income of $0.50 to see everyone else gets no more than $0.50.

4. Unevenness on wealth distribution arouses hatred, and the Socialists' vow to bring the evenness to the reality for everyone by taking advantage of such hatred. To make "evenness" happen, however, Socialists say they need to gather power from the masses.

5. Isn't it nice for those who get $2 today to get the income of $10 tomorrow? It is even nicer to see those with income of $100 or $10 today to suffer tomorrow. The Socialists are not only right, but they also give hope.

6. So, Socialists: I contribute my power to you, you guarantee me a better tomorrow, OK?" It looks like a lucrative gambling without money bet required.

7. As the gambling is collecting bet, someone also smells a business from which profit can be extracted without risk in democratic society: to drumbeat for the Socialist agenda. The profit is in many folds: to win a wreath of nobility by showing sympathy to the poor, to be able to secure a market for their opinions, to prequalify themselves as someone whom the future power must rely on. The only investment is to more intensively fuel the hatred caused by the unevenness of wealth distribution that everyone sees.

8. In contrast to the full exposure of unevenness of wealth distribution in democratic society, Socialists guard everything of theirs, including lifestyle, behind a secret door. Any attempt of investigation is a grave crime. During the Yan-An era, a young Communist, who was actually full of naïve democratic idealism, died because he pioneered for a group who intended to expose the lifestyle of the high leaders in the Chinese Communist Party. After his death and ever since then, everything in that direction was forever quiet. With these contrastingly different operations, hatred caused by unevenness of wealth distribution only inflames in democratic society, but Socialist society is forever conceived as beautiful as the fairness pursuers would purport. At least, no unevenness of wealth distribution is detected in the Socialist society, although, for example, there are twenty-six ruling classes tiered with different power and privileges behind door above ordinary people in Communist China.

The fairness pursuers in America seem having not yet presented a Socialist agenda that is as radical as what Vladimir Lenin and Che Guevara had ever presented. However, by declaring to work on removing unfairness, they certainly have been able to present unwritten laws to overrule almost any written laws in America. Example is shown by the stand they take in the conflict between America and her enemies. When the American soldiers are caught by enemy and tortured with any extent of cruelty, they are never heard of saying one word in protesting. When they learn reports about an enemy combatant being captured and tortured by the U.S. forces, they would show all the heroism, vowing to have related personnel brought to court to face justice. When asked why they are found showing such dramatic difference, the answer is this: "The enemies did do this [torture] on us, but we cannot do the same thing on them; we cannot play at their level." Or "each captive has alienable right" is another answer. Brilliant! Their answers show how they have dictated who should receive the justice of the law: only those who guard the nation, not those who attack her.

A white anchorman must lose his job because he made a comment discomforting some nonwhites; a male university chancellor must lose his job because he made a comment discomforting some females. However, a nonwhite female's comment in which males from whites should be suffering must be made immune from any negative evaluation on her way toward a highly revered office. The point here is not to discuss which comment is right or wrong, the point is why the fairness pursuers think what they are doing is fair. Losing a job is losing a lifeline to many people. If it is not a dictatorship discriminating against whites and males, worse yet, against the combination of a white and a male, what is this?

—

106

Are more examples needed to illustrate how the essential Socialist operations are matched: to remove capital monopoly but to install monopoly of everything, and by the same token, to remove discrimination against a few but to install discrimination against more? All these are done under the umbrella of democracy, but all are done on the way to bury the genuine democracy.

Chapter 8

Democracy and Human Right

a. Democracy, Nation, and Competence

Kindness or unkindness will not pronouncedly determine a government's continuous existence; it is the competence or incompetence determining her continuous existence. Which benevolent government without competence has been known long lasting in human history?

With the benevolence toward her overall populace and the abundance on material life, America historically attracts to this land incessant flow of immigrants at all costs, legal or illegal. However, with her piling up incompetence in the political world, she is seen being pushed by domestic as well as foreign forces to the verge of struggling for continuous survival. Evidences? Look:

Twin Towers on her land can be wiped out in the blink of eyes by plots organized by a small group half a globe away; more than twelve million of foreign personnel, whatever they are called, can flood on her land, waving flags of other country in demonstration as if this country had been conquered; it is almost a daily routine news that students, teachers, law enforcement agents are butchered as if in a game hunting range; her manufacturing lines have been seen not too many left at home; as a liberator of 150 years ago, she seats there with absolute lunatic when the descendants of the liberated consider her as an equal of the disappeared government . . . Ironically, people do find her competence in another field: she has taken only 30 years to convert Communist China to her biggest creditor, while this country was dirt poor compared to Uncle Sam 30 years ago; and the U.S. now intends to demonstrate to the world how she can manage the biggest national deficit in human history without too many manufacturing lines on her land.

Despair of the nation's future is heard every day on some radio talk show. Will a nice uncle like Sam survive long with the incurable incompetence? What can be detected is that many international factions are predicting the looming bankruptcy of America; many of them have seen preparing policy to leap in for the kill. A test balloon is released already: a very recent suggestion aiming at replacing the American dollar with a world currency is heard from China and other countries. Even to a not-so-strong country like Mexico, event after event has indicated that nothing smaller than a big chunk of land out of California, Arizona, New Mexico, and Texas will satisfy her craving stomach.

With her benevolence to her people, America is no doubt the most well-known democratic country in the world. On the other hand, the Communist power in China has never been considered benevolent to people under her dominance; the record involvement of life perishing in her entire history since her birth day of 1921 is well known. Yet whether one would like to accept or not, Chinese Communist Party is a very competent political organization, and the government under her dominance is very competent in ruling, regardless how the people feel they have suffered under her control. It is said that the inflation rate soared to 1,300 percent in her military base during the civil war. With this economically destructive figure, however, the Communist power pushed the former dominant power to Taiwan. Between the years of 1959 and 1962, it had the overall famine that even Africa may not find a match. Record showed that cannibalism, even parents eating children, appeared in some areas; parasite from human discharge, if found, were even considered food source. However, during those years, China was able to manage herself to be militarily untouchable by any outside forces; of course, it was absolutely peaceful inside too. During the years from 1966 to 1976, the so-called Cultural Revolution turned the entire country upside down; even the later Communist leaders documented this political event as the biggest national cataclysm in Chinese history. Dire as it was, the Chinese Communist government managed to have two big international events to happen in her favor in those years: (1) the American president devoted a unidirectional effort to visit her, a country with which America has no diplomatic relationship; (2) the American military forces must retreat from the Vietnam battle ground, although the United States seemingly succeeded in blocking the Communist scheme from dyeing the entire Southeast Asia red.

All these comparisons serve to make us be aware of one thing: It is not how good a government is to her people makes her last; it is her competence to make her last. It is an ideal government if she could be benevolent and competent at the same time. If she could not present both at the same time, how would people choose? Oh yes, they would choose the benevolent one, but at the same time, make it incompetent and even destroy it too. With the abundant freedom and material life that America allows her people to enjoy, it can be said that anyone who finds himself living intolerably in America can find nowhere else to live in the world. Yet American government must find herself confronting with the most populace with complaint, or even hostility. Why?

b. Excessive Democracy

Democracy, a powerful element that ferments the upheaval of Western culture, has been pushed into a state of excessiveness, one of the sicknesses that the Western culture seems unable to cure. What is the borderline beyond which democracy is considered excessive? There is no good answer to this question. However, a society of excessive democracy must be associated with a political system, a legal system, and a moral system in which no authority can be consider irremovable. Let us explore how some excessive democracy syndromes have been functioning in the West, typically in America.

Removal or nearly removal of death penalty. The West mistakenly thought her culture is ideal while it is not, and can never be; her people falsely dream that they live in an ideal society, while they don't, and can never do so. By removing death penalty, however, the Western people merely naively switch the power of execution from the hand of a government to the hands of the potential criminals, but then encourage more death. In some sense, such hand switching in killing is not only irresponsible, but also practically makes the government a cold-blood cooperator of the criminal. At least, in the tyranny government, there may exist some predictable "guidelines" for people not to touch, otherwise death penalty would follow. In the democratic society where death penalty is removed, however, there is no such guideline given by the criminal on the street; so they kill as they like.

Abusive allowance of the use of insane to allow criminals who deserve capital punishment to escape death penalty.

Law enforcement agents easily got framed. The West, particularly America, has no lack of example cases that a law enforcement agent is easily sent to jail by someone who should have been the one to be jailed. This practice can only lead to an idea that the government views a criminal's life with higher value than that of a law enforcement agent.

America has never been short of examples of his people, from top leaders to the grass root masses, siding with ideas advocated by the enemies of this country. During Vietnam War, a famous U.S. movie star was found advocating anti-American ideas in Vietnam. During a war against Taliban, an American young man was found being a soldier of the enemy. The nation's interest has no weight in these people's mind.

The low cost of committing crime, coupled with other reasons, makes the U.S. national secret data bank, military or commercial, a public library of 24/7. Espionage against America is an almost-worry-free business for someone to earn double income; benefit can be realized from both success and failure in such an espionage activity. One man was charged of being a spy for the Chinese government, and his lawyer threatened to have more secrets to expose in court.

This man not only later received an apology from the government, but also a lucrative compensation from the U.S. government for his "suffering."

Privacy of individual is seen having higher priority over national security. Even phone tapping on terrorists is an uphill struggle for the American government security agents. Many American citizens have a naive attitude toward democracy: democracy is not genuine until all means protecting democracy is clipped off.

National adhesion should have no authoritative value if it hinders democracy. For such democracy, English is not allowed to be documented as an official language, while other languages, such as Spanish, has implicitly shammed in with the status of semiofficial language in practice. Could a land become a thriving country that is called the United States of America for more than two hundred years if this land had started with thirteen languages at the beginning? It does not matter. English has finished her historical duty; now it is time to dismantle it for "democracy."

Illegal pursuers of American dream are tightly protected with much higher cost than the national sovereignty, free medical caring, welfare, social security, free education for them even makes the out of state students holding American citizenship envy, oh, also a big budget preparing for their well being in the jail if some of them happen to be unlucky. "Let them go, they only come here for an American dream, they don't hurt anyone," shouted the democracy promoters. The same promoters, if happening to own a house in Beverly Hill, California, usually also have a sign Gun Response posted on the property line as if the national borderline had stopped only a few feet outside the property lines; and behind those lines, there is no American dream to pursue.

Any of the above would make a society seriously suffer, but the West, particularly America, have embraced them all. This further tells people that democracy is not only incompetent in self-defense, but also that she could not make her people happy until she completes her journey on a suicidal path along which more democracy has to be produced.

A big reason that the West allows excessive democracy to flood is the longtime easiness in life that has made many people lose sense in distinguishing personal conduct from governmental behavior. Such confusion leads them to believe that a government must be wicked unless its behavior is a reflection of nobility seen in personal conducts. Subsequently, political arenas are imagined as rings of perfection in which nobility is to compete. Therefore, sharing, regardless with whom, is advocated; but fearsome force from the government function, regardless against whom, must be suppressed or even removed whenever possible. To these people, the top priority in life is to fight for right; personal obligation to a nation is seen as an intolerable oppression from a wicked government. Human rights, devil rights, all rights—regardless how ambiguous their definitions are—as soon as they are termed as "rights," are made invincible axes toward all authorities. Whatever seems impenetrable by law, just use various names of right to crack it open. While this trend of ideas flooding, various political factions with different ambitions see their opportunities too.

c. Human Right Serving Whom, Devils?

Among all so-called rights, human right is the most ambiguous one; in the name of human right many political factions are able to push democracy to any extent of excessiveness. The number 1 ambiguity is that it only stresses the protection, regardless who is qualified for protection. So more often than not, European Convention on Human Rights (ECHR), for example, guarantees a death-immunization ticket for someone who has taken away the life of another person. So with her blindness on protection qualification, ECHR just guarantees two individuals with different values of life between a murderer and an innocent victim. Obviously, the criminal one has more right to continue his life because of ECHR. The number 2 ambiguity is that it has left an unlimited blank space for other desires of human beings to fill in as human rights. Should any human desire not covered by the article be included or excluded as human right? For example, how is enjoying marijuana not a human right in Amsterdam, Netherlands? But in the name of human right, why should such an enjoyment be locally limited somewhere else? How is homosexual activity not viewed as an alienable right? Between abortion and antiabortion, which is human right? The number 3 ambiguity, and the most fatal one to the legal existence of the West, is that it has guaranteed to limit the ruling power of a government, but provides no limit to the attacking power of the organization that are hostile to the government. In other words, those expecting protection from the government are exposed to more vulnerability and more suffering than those aiming at destroying them.

With all its present ambiguities, more often than not, the realization of human right is at the expense of the corresponding right of some other group of people. The cruelty it brings up in many cases is even more hideous to some people than this right has never been an issue. Instead of protecting people as what ECHR claims, it is always used as a big flag by some special interest groups to cover up their agendas that they otherwise cannot push forward. Let us discuss more in details about these ambiguities.

Because of ambiguity 1—i.e., no qualification about who is protected—the following ridiculous thing happens: an inmate sentenced to death is entitled to a kidney transplant, while the family suffering from the loss of a loved one must continue to pay tax for the hilarious expense. How is the human right for the inmate not fully realized if he is given a choice how to die: either going to the gas chamber right away without further suffering from his kidney problem, or keep his kidney problem until his designated death day? In some murdering case, some victim was repeatedly stabbed or tortured before death, but then the murderer is closely watched to make sure he will receive the least suffering punishment. Is the ECHR set up to protect criminal or innocent people? How would a law have become a more civilized law only because the life suffering from death is an innocent one but the life enjoying protection is a criminal one? How would a law have become a more civilized law only because it enables a criminal hand to kill, but restricting a justice hand from killing a criminal? How would a law have become a more civilized law only because it would further victimize the victim's family to pay the price in protection of their destroyer? How would a law have become a more civilized law only because the fearsome force from a government is taken away from a criminal's calculation before he commits the murdering? Not fully accepting ECHR, Communist China put 1,700 criminals in death out of a population of 1.3 billion in 2008 (http://asiadeathpenalty.blogspot.com/2009/08/china-demand-for-clemency.html, Oct 28, 2009); while fully accepting ECHR, American had more than 10,000 innocent lives lost to criminals yearly out of a population that just barely 300 million. Isn't it a clear picture whether ECHR is there to protect innocent people or to encourage criminals?

Ambiguity 2 of human right is its failure in rendering exact definition of human right. True, ECHR does cover some human rights, but it fails to point out what exactly should be include or excluded as human right. Subsequently, many human's flesh desires are able to be wrapped with human right decoration, and pushed forward by some people to fit their agenda, forcing another group of people to accept. The declination of such acceptance can very well be declared as anti-human right, an implicit crime even before any legal trial is needed to begin. Some typical issues in the name of human right in nowadays society, besides the elimination of death penalty, are listed in the few upcoming paragraphs. Through these paragraphs, people can tell that all these proposed human rights are demanded to be recognized on the basis of destroying an equivalent right of the other group of people. To make it even more horrendous, in most of the cases, the group to be guaranteed to receive benefit has far less popularity than the group to be destroyed.

1. *Abortion or antiabortion, together with the so-called feminist right and the man's right.* If one side is defended by human right, the other side must be declined, but which side should be defended?
2. *Premarital birth or, more directly, unwedded birth.* Should mothers nurturing premarital birth be viewed as being irresponsible and be penalized, or should they be sympathized, even be praised in some cases, and rewardingly helped? That a welfare system supporting these

unwedded mother and children can exist is because some responsibly hard working people have been taxed by the government. How the unalienable right of liberty of a group of people has not been violated when they must be forced to pay tax for the private joy of another group of people? Isn't it the same nature in the slavery years that the slaves are made by law to pay for the private joy of the slave owners?

3. *Homosexual activity.* Under the flag of human right, homosexual communities show extraordinary bravery in challenging all the authorities that are traditionally rendered by morality and legal practice in the West society. "What is wrong with homosexuality?" asked the homosexual lovers when confronted with people who hold disagreement on homosexuality. So far, however, they are not found presenting statement on "what is right with homosexuality" to the society. When the homosexual lovers insist the society to create a new definition of right for them, have they shown how much human right they have reserved for the others? Before "what is right with homosexuality" can be answered, before feeding back of equivalent benefit from them to the traditional community can be made acknowledged, the following remain puzzled:

    a. Could the bravery shown by the homosexual group in the West stand alone without the protection at the expense of the group of people of heterosexual practice? It is obvious that heterosexual lovers can stand alone at the absolute absence of homosexual lovers. If one group of any living beings can continue the existence without the other, and the opposite is not true, one of them must be a parasite in some form on the other group. Of course, in this sense, very young children may be viewed as parasites to their parents too. However, children offer hope to their parents. On their shoulder, there is a big task: to extend the biological significance of the family to the future. This is a fair trade. Is there any fair trade existing between the homosexual group and the heterosexual group?

    b. With the barren nature of homosexuality, why is an activity discontinuing the human being supply to the society is a human right? Is the concept of human right proposed to flourish the human society or to extinguish it?

    c. At the age around fifty, heterosexual lovers may have to suffer some heartbroken news: their son was killed in a war defending the nation. Does any homosexual lover have to experience the horror of a child loss? What kind of right can they offer to the parent and the dead son for compensation? Do not cover this question with child adoption, as the homosexual lovers provide no their own resources.

d.   As the son soldier dies, all money and time that the parent spent on the son during his growth is gone; they all are contributed to the nation's well-being. *The most precious treasure from the family is given to the nation free!* During the same time of the son's growing, what do homosexual lovers do to contribute to the nation's future with the same magnitude? How have the homosexual lovers contributed to the nation while enjoying the benefit provided under the newly created category of human right?

e.   If homosexuality must be regarded as human right, how should heterosexuality be regarded? A child adopted by homosexual family is forced to be trained and educated in a homosexual environment. Even the homosexual community admits that most people grow up with heterosexual preference. How has this adopted child's human right not been deprived by being forced to grow in an environment that most children do not have to be in? By probability, how does the homosexual family determine the adopted child possessing no heterosexual preference? If unable to determine, how has this adopted child's human right not been deprived if his heterosexual preference must be suppressed?

4.   *Problem of multiple languages, typically in America.* If any other language can sham in to gain the status of semi official language in the name of human right, where is the human right for those English-speaking people? English-speaking people not only have human right but also lawful right to insist English to be the only official language in America; why is their right seen constantly rejected? As the Spanish language expands beyond any limit, people hear more and more often such statement: "It is what is going to happen. If they [English-speaking people] don't like it, they can go back to Europe." It is a statement openly buries the human right of the English speaking people.

5.   *Illegal borderline intruders.* It is said that it is inhuman if welfare system does not cover the illegal intruders, and that it is inhuman if a parent with illegal residence is deported so that a child is left behind, and that it is inhuman if a language barrier they show no willingness to overcome is not removed . . . However, human right promoters on behalf of these intruders never answer one question: How has the human right been better protected for the group of citizens who feel so threatened as they see the sovereignty of the country vanishing? Is this what it means by human right protection, in which one group of people of a far more

---

population must have their human right trampled? Honestly, resources, even fresh water, for each country is limited. Why are the citizens not allowed to guard their resources but forced to share with someone only because this someone has bravely but illegally appeared in this country? In this sense, why are the colonists three or four centuries ago seen so criminal?

6. *Racism in new direction.* This problem will receive a thorough discussion in other chapters and will not be elaborated here.

Ambiguity 3 of human right is its grave effect in limiting the government's ruling power. A democratic government is supposed to protect her absolute majority of people. If the ruling power of the government against a few thugs cannot be fully function because of the so-called human right, how good is this government to the most people in the society? How is this government different from an irresponsible government? Is the democracy set up for most of the people or for the few thugs? The protection offered by the ECHR is often seen operating in this way: The enemy dangles in TV a head chopped off from an innocent person; protest from any justice crowd is seldom heard. However, wave after wave of demonstration against the government shows up in the street if the government is said to have "improperly" acted upon the hostile combatants. So while ECHR fails to act to have some innocent people to return home, it surely has enabled the detainees in Guantanamo to return to their former bunkers with smile, declaring they are terrors to the bone. As the detainees in Guantanamo are pressed to be released because of ECHR, where are the human rights of those innocents who were slaughtered in a big scale within a short time period in the Twin Towers in New York? Where is the human right of the victims of suicide bombers?

In short, where is the human right for those people who demand from their government the immunization of fear? Why do their enemies have human right to see no fear from the same government when they so sternly decide to destroy the innocents? For what purpose is the government not allowed being equipped with power to make the enemy feel fear?

The Western democratic society, particularly in America, has gotten used to such a mentality that human right is generally found violated, or at least infringed, if concession is not made to produce by the first group of people in each of the conflicted pair in the following list:

1. Parents vs. children
2. Teachers vs. students
3. Personnel against drugs vs. personnel for drugs
4. Users of primary language vs. other mother tongue speakers

5. Local residents vs. illegal borderline intruders
6. Heterosexual lovers' community vs. homosexual lovers' community
7. Law enforcement agents vs. criminals
8. Victims vs. destroyers
9. Creditors vs. debtors
10. Obligation toward a nation vs. rights of a citizen
11. National security vs. privacy
12. Morality vs. sense satisfaction,
13. Standard in competition vs. desire of no standard in the same competition
14. People with more career success vs. people with more irresponsible behaviors
15. Christianity belief vs. beliefs not in agreement with Christianity,
16. Brutal force defending a nation vs. brutal force destructive to the nation
17. White vs. nonwhites
18. Overall, a group asserted having a stronger background vs. a group asserted weaker

Why must people in the first group of each conflicting pair be damaged for the desire of the second group in order to see human right to be realized? Aren't people in the first group also human beings (this question has been as ridiculous as it can get)? And do they have right? In the American history, which group in all these conflicting pairs has been inspiring America more to reach her prosperity? Let things carry to the extreme, which group in these conflicting pairs will preserve America's prosperity better? Or which group will add more weight to the American vessel that has been shown deteriorating?

While democracy has a noble idea not to leave people of minor group behind, it is certainly a wrong idea to satisfy the minority group by damaging the majority group through government power in order to get fairness. This is the same nature as a tyranny society except it is even worse as far as self-sustenance of a society is concerned. In an openly stipulated tyranny society, the ruling is from top to bottom; in the deformed society with excessive democracy, the ruling is upside down but through some government agents. In an openly stipulated tyranny society, the ruling class mobiles the entire society with more overall intelligence from a bird's-eye view, in the deformed society with excessive democracy, the ruling class just bring more chaos to the society from the bottom, and more often than not, with a lower grade of intelligence. Simply imagine a team of climbers in action on a steep cliff: a stronger one in front always looks back and lends a hand to someone weaker behind, who struggles but still tries all his effort, finally they both reach the summit. However, if the weaker keeps complaining and stops climbing but only tights himself to the waist of the stronger one, but somehow someone else can force the stronger one to keep climbing, what kind of social "cooperation" is this? And will they both be able to reach the summit or finally both fail?

The social strength that the aforementioned second group shows in overpowering the first group is no doubt invincible most of the time in spite of their smaller population. Therefore, people could have confidence in believing that someone is investing with a well planned agenda. For this agenda to be pushed forward, human right has been clearly serving as a flag leading the second group of people to march from one victory to another, regardless what is the actual content of their demanding under that flag.

The excessive democracy, led all the way by human right, has shown more and more grave effect in jeopardizing the competence of all Western governments, particularly the American one. The incompetence of the U.S. government is so overwhelming that people, no matter how they have labeled themselves—Republican or Democratic or Independent or conservative or liberal—all see the government not doing the job for them. America, this Gulliver, has been tied down by too many ropes tagged with human rights, and all these ropes started from the thickness of a silklike thread.

While working extremely effectively in tying down the Western governments, has the ECHR ever had any effect on any country or any political organization that can be seen in violation of human right? Practically none! Soviet Union had never given it a look during her existence; North Koreans people are not allowed to know of her existence. The Chinese Communist government, knowing that the foreign capitals would otherwise hesitate to come in, showed gesture of signing some document for her acceptance. However, she has been able to retain full freedom regarding how much she allows her acceptance to become valid. Organizers of suicide bombers have never received one word of condemnation from this ECHR, while some people in some authoritative position thought they should and could extradite a retired American vice president according to the same ECHR. By giving different measure between government and the destructive forces against the government, ECHR has set up a norm in the Western society that policemen deserve no honorable respect but hatred.

Setting aside how damaging the ECHR has been in "protecting" the democratic society, its efficiency has enough to raise concern. Without the ECHR, the democratic world is able to bring justice to the major war criminals of the WWII in the Palace of Justice, Nuremburg in eighteen months after the war stops. With it, however, some similar trial stand is still not seen able to host anyone who must be held accountable for the decimation of one quarter of the Cambodian population, even after those lives have perished for more than one quarter of a century. ECHR has become a platform for the democratic society to demonstrate to the world how nobly she could kill herself and then to dream of that the thugs of the world would follow her graceful example to meet their end.

It is understandable why a document like ECHR seemed necessary at a certain history background. However, after the practice of more than half of a century, it has proved that it only causes detriment but nothing else to the Western society. On the contrary, forces that are destructive to the Western society have received full benefit from this document. If the West civilization wants to extend her breath, it is time for her to retire this document, or fundamentally revise it. After the revision, it must precisely emphasize what are to be included and excluded as human rights. It must also be able to declare who cannot receive its protection. If there exists no such possibility to have all these conveyed in the document, its absence will serve the West better than its presence.

Democratic society must stop the practice in which one group, usually with smaller population, enjoys human right at the expense of the corresponding human right of the other group, which is usually a much bigger group.

Chapter 9

Racial Evolution

a. Different Concern

Culture in the present world has evolved into a stage in which values that are once pioneered by Caucasians show sign of retreating in all aspects. The utmost important reason for this to happen is the thinning of their population and subsequently the overall shrinking of their habitats. This must result in their gradual but rapid fading away of their dominant influence. Situation has been so overwhelming that, in some of their original habitats, a number of newborn babies of non-Caucasians begin to surpass that of the Caucasians. To further reinforce their difficulty is that, with the wealth they command, a big portion of them takes the values of a corruptive lifestyle and excessive democracy. Corruptive lifestyle erodes will; excessive democracy helps to secure a space where eroding can go on with undisturbed comfort. To maintain the enjoyment, they found that minimizing the number of offsprings is the most effective way in reaching the aim. Besides being highly watchful in birth control, the following strategies also pronouncedly contribute to thinning their population:

1. Their women are encouraged to compete with men in all fields. From the point of view of human social value, there is nothing wrong with it. However, does Mother Nature look at the matter in the same way? Highly educated, their women vigorously seek for power in all fields, rearing younger generation is found as a big hurdle for their ambitions. To overcome the hurdle, they choose to minimize the pregnancy, or postpone baby bearing to a much later age, or even completely avoid conception. To any species of animal, if their females' behave coincides with such strategy in lineage continuation, it will not take too much mathematics to predict the end.
2. Homosexual activity is highly encouraged under the cover of human right. At the level of personal conduct, anyone could take care of the genital portion of his/her body in any way he/she wants as long as no

one else is victimized in such care taking. However, does Mother Nature look at the matter in the same way in her playground? An obvious result of homosexual activity is its barren consequence. Some may argue that homosexuality is a universal phenomenon in mammals, why would it only render negative results to Caucasians? Very simple, for example, Islam forbids homosexual. As the production line of newer generation of Caucasians is constantly curbed with powerful brakes while the Islamic line is prolific in full speed, does it need a genius to predict what the outcome would be? Indeed, it can even be suspected that someone has been investing in rewarding the homosexual activity in the West, particularly in America, from monetary benefit, to power (temporarily, of course), to honor. Can any gun or atomic bomb surpass this effortless investment in matching the investor's long-term interest, spanning from dethroning the whites from the office of power because of their thinning population to having them to silently relinquish their land?

3. More often than not, their marriage and divorce system ends at the heavy economic damage to the males. An unhappy ending of marriage often becomes a long-term heavy economic liability for a male. This must result in halting the males' willingness to commit to their females, and number of formation of family must subsequently suffer. Why should one commit when sex is almost free and of no consequence of any kind with the use of condom? On this opinion, their females must disagree, but the males must have their own formula for calculation too.

4. Even being more dreadfully threatening to their population sustaining, at the same time they curtail the forthcoming of their own offsprings, they pay to support a social economic system that other races' prolific reproduction can take advantage without bound. Who said that baby making with knowledge that financial support is automatic is not one of the biggest joys in the world?

5. This may be trivial, but it happens: white couples adopt babies from other races, but white babies being adopted by other races are undetectable.

All the above reproduction arrangements are only elements adding weight to the other elements that have long caused the whites' population to sink. For example,

1. Caucasians have never been able to form a stable nucleus for their gene group. In history, at least in the Western part of Europe, they have allowed the slipping away of many chances of forming a large country of a more homogeneous population. Modern political leaders such as Napoleon and Hitler did launched military action to unify Europe, but

both committed grave mistakes and failed. So the historically split gene group stays split. To make it worse, wars, one bigger than the other, keep combing back and forth on this land. In European history, one can hardly find a period of fifty years during which war is not found.

2. Not only has their gene core been decentralized, but most of their habitats have also been highly permeated by other races except few, such as Russia. Compared to the Caucasians, blacks, Mongoloids, and Middle East people have been keeping concentrated gene nucleus within stable habitats.

3. Within the same gene group, they have far more languages, not even dialects, than they have countries. History has given us enough lessons to believe that diverse in language is a number one killer to the national adhesion. To effectively disintegrate an ethnic group or a country is to first ruin her language.

4. There is no uniform belief. Seemingly dominating, Christianity in the world of belief in Europe is branched into three large branches: the Roman Catholic, the Eastern Orthodox, and the Protestant, not to mention some more further diverging denominations under them. Among all these branches and denominations and diverging lines, sometimes they see each other even more intolerable than believers from other religions. Aside from all these, Islam has a long historical root in the community of Caucasian too. Besides beliefs devoted to supernaturalism, forget not other beliefs such as atheism and Communism.

5. The Bolshevik movement and the two world wars gravely began their population decimation.

6. Their high level education but a decimated population imposed on them the feeling of a need to import cheap labor. This is their most nightmarish element of succumbing because they begin to lose habitat to the troop of cheap labor. Through importing cheap labor, they have created for themselves a bondage, which requires money payment, a welfare system, and land to satisfy. Any human wealth can be created by man, but absolutely not land, on which habitats are based.

Can the white as a race be saved? The chance would be almost as good as the chance that all the above points can be reversed. If a fundamental operation for this race is not overcome, if they cannot organize on some base that pays particular attention to the welfare of their race, any talk about saving the white is a waste of breath. The fundamental operation is their volunteering reduction on their population. If they do not save themselves, nobody can. At one time in a Fox interview, a basketball celebrity told the program host with pride that his family had eight siblings, all being raised by a single parent. Such an able single parent must be very resourceful for some reason and is not heard of among many working couples. Even with both parents in a family, many working couples would have felt their ability stretched to back broken extreme with three or four kids. Setting other elements aside, population contest alone will have the whites doomed.

But why save them? Oswald Spengler is always seen as a negative culprit in the eyes of many antiracists. Any idea suggesting saving the white is condemnable, not to mention if such an idea can be practical or even executable. Indeed, some of the ideas contributing to saving them are hated even by a big portion of whites. For example, would any one expect to encounter no resistance in proposing the following suggestions?

1. Stop homosexual activity.
2. No drug abuse, not even marijuana. Nearly two centuries ago, the British invaders effectively paralyzed the Chinese with opium when they resisted the aggression of the British. Now, a big portion of white seeks to get paralyzed with even more effective agents, such as heroin, cocaine, etc.
3. Each white couple must produce minimum number—say, four—of children within a certain years of marriage by law, otherwise heavily taxed.
4. White woman's refusal minimum childbearing before a certain age is regarded as serious a crime as man's refusal to be drafted by military force. No job of any significant pay in the public sector or private sector can be offered to woman unless she has met the minimal requirement in childbearing.
5. European Union will allow only one official language. Oh, yes, the entire European population will rather "die of honor" than seeing this to become a reality, until their honor is taken away. Ask a French person to speak English? Why should he? He must reserve the pride of his mother tongue until, well, until he sees Berbers having an irreversible potential to replace the French.

It is hopeless to expect anything similar to the above ideas to be carried out in America. Whites can only restore themselves in Europe, Eastern or Western, before their hope is completely gone. However, at least, in America, equal status and protection on legality and morality should be given to organizations based on Caucasians heritage or European geographical ethnic background. If the same can be given to organizations based on other backgrounds, such as black, Latino, and Pacific Asian but declined to be given to Caucasians, it is an obvious racist policy. Why a race that shed so much blood in liberating color slaves in American Civil War is deprived of the opportunity to enjoy the same right as the other races have enjoyed? Why a practice is not racism if in this practice whites are selectively left out or even forbidden but everybody else is encouraged? How does this practice differ from the old one in which houses in a certain areas were stipulated not to be sold to nonwhite people? Is installing racism but in another direction a desirable way to remove racism?

A fifteen-year-old California white girl was forced to transfer to other school after it was learned that she planned to start organizing a Caucasian club. She was smashed from everywhere with scolding of being a racist. She is not seen as having right to do this; no protection is offered to her from any one. This is horrendous. At transferring, she must feel heartbroken in believing the justice that her ancestors once joined with others to struggle for every citizen. On this matter, all human right warriors were found absent, leaving her human right being maximally trampled. The government and the school watched silently, allowing the hatred motivated by obvious racial reason to pocket in the victory.

One and a half century ago, "nigger" is a term for the white racists to pass a message to the blacks: "Don't even try to get what I get, we are not equal." Today, in the same country, "racist" is a term for few nonwhites to pass the same message to the whites: "Don't even try to get what I get, we are not equal." Indeed, the use of "racist" is ten times more forcefully in straightjacketing a person than "nigger." "Nigger" is only seen to be skin wrong by someone's opinion, but the term itself bears no moral or legal measurement; a racist can be crucified by law. Framing a person with the term "racist" enjoys the convenience of "lawfully" depriving such a person from something without a trial.

Pictures are chilly to the whites, not that much to the current whites yet, but to their future generations when their population is seen to be thinned to a threshold number. Even so, witch hunt already begins way before they truly fade away. In nowadays society, while nonwhite is untouchable in public opinions, one can attach negative value to Caucasians as freely as he wants. Didn't lately we hear a complaint that blue eyes and blond hair being the culprits who sink the entire world to the current financial crises? But are the Caucasians ever credited with the prosperity without which financial crisis has no background to show up? The free pinpointed witch hunt is not a good sign for the Caucasians.

As history has evolved to as is, the chance is bigger than not for the whites to watch themselves further paralyzed. Yet in front of their racial crisis, so many of them are still fond of flesh enjoyment, or hedonism, which can make people willing to pay any price for it, including death. Possibly many people have been too familiar with the story how Anthony lost a decisive battle to Octavia during the Roman era because of his loss of will at the separation of Cleopatra. If someone is willing to pay anything for temporary comfort or enjoyment, no one should expect him to think too much of the so-called future, or even future generation. In America, unless Christians can restore their "home ownership" responsibility in the nation that they are expected and obliged according to the Declaration of Independence, America has no hope to expect her people to think of the future, the future generations, and the future of the nation. If it has been time to mention about population browning of the world, wouldn't it also be time for curators in some museum of human evolution history to prepare a new skeleton stand next to the Neanderthal one? A statement next to the stand describing the disappearance of Neanderthal may not have been more accurately applied until now: the appearance of some more intelligent *Homo sapiens* have pushed this race to the exit of unfit.

With a yearly loss of seven hundred thousand people on population, a Caucasian country like Russia is not focusing on how to increase their population but to restore their past glory of Communist dynasty. This means that they still think they can afford to join more games in which more blood and flesh from their race are sent for disappearance. Can a country really make the world tremble under her feet without enough people? Today, only if she can rearrange her priority in applying her resources, converting from an anti-West strategy to nurturing a prolific population, then would she revive the past Russian glory, not as a Communist monster but as a thriving country that had a long history of profound culture, a culture from which names like Peter the Great, Tolstoy, Popov, Mendeleyev, Pushkin, and Tchaikovsky sprang up. Otherwise, sooner or later she must find looming of such a day that she could no longer defend her extensively long borderline, not to mention any glory. General Kutuzov could do nothing when he found he had not enough soldiers to command.

When Europe is succumbed to the prolific Islamic population, when the whites are successfully overshadowed by the nonwhites whose population expands at exploding speed in North America, when Asia feels the increasingly mounting population pressure but carries a swelling wallet, Russia' share of land per capita must be under close scrutiny for comparison in the world. With one eight of the world's land being occupied by 2 percent (and in the process of getting less and less) of the world population, an outcry of unfairness is easily formulated. Didn't some Russian leaders not too long ago criticize America as being unfair by consuming one-fourth of the world energy production while she has only 5 percent of the world population? What makes them to expect that other people will not find Russia as an unfair culprit on something else? As a rule of thumb, people proposing criticism of unfairness will not usually limit them to criticism. It is definitely puzzled to learn that someone decides to have some wildcats to patrol around his house with the hope that an old domestic cat of a far away neighbor will thus be stopped from entering his house to soil his carpet.

Will Caucasians just appear like a meteor in the night sky in the long history of human being evolution? Historically, until other evidence turns up to tell otherwise, the oldest skeletons that are found closely related to their ancestors are just the forty-thousand-year-old Cro-Magnon. How is this age compared to the age of the seven-hundred-thousand-year-old Peking man or that of the 1.5-million-year-old *Homo ergaster*? Civilizationwise, Caucasians do have radiated dazzling brilliance along their evolution journey in the human history: science, art, medicine, philosophy, discovery, invention. However, from Cro-Magnon man to how they shined and shook the world, to the present dwindling population, all seemed to be happening in only one eye blinking of history. Is this God's purpose to push them in a hurry? There is no doubt that worldwide replacement of Caucasians by others matches ambitions and interest of many groups of people. At least, Caucasians are not a favorable element to be seen around by someone who finds it more desirable to have landscape and real estate value matching those found in other continents, such as Africa. How will a world look like without Caucasians?

Human beings should be still on Earth without Caucasians, but the balance in every phase of human life will not be the same without Caucasians, a race that has been pioneering a culture that has been so widely accepted for so long. There is no doubt that all races are able people. No evidence has suggested that people who left behind with a brilliant culture in ancient Egypt were Caucasians. China was once a country of the highest GDP in the world in history. Indeed, after the Chinese wake up from the Socialist nightmare (neither completely nor willfully, though), it takes only thirty years for everybody in the world to expect that the gravity center of the world is shifting toward them. Japanese did show that they were capable of inflicting military wound upon some Caucasian countries that they still lick with painful memory. The glare from the historical sites of Maya and Inca cultures can still blind people's eyes. But given the high frequency of historical association of invention, creation, and discovery with the habitats of Caucasians, should we believe that science and technology development still take the same pace along the same direction when Mother Nature must push the Caucasians aside? Imagine in one minute that if the medical practice and methodology started among them and fundamentally perfected by them must be removed, do people have any equivalently effective replacement to keep themselves healthy? Of course, someone can still formulate a complaint on this medical advancement: In these one hundred years, world population has increased by almost four times. Human population pressure has never been escalated to so intolerable at such a speed in history.

—

b. New Racism

What is *racism*? Copied from a dictionary, a definition is given as the following:

1. The belief that race accounts for differences in human character or ability and that a particular race is superior to others
2. Discrimination or prejudice based on race

According to this definition, racism can be an unreasonable political pressure from any race. However, the current society seems to be accustomed to a mentality that only unreasonable pressure exerted from white to other race is termed as *racism*, but that the same unreasonable pressure from any other race to white is not racism and should be taken for granted.

Given the weight that America had shown in the world, given the diversity of races shown in this country, a peek at the conflict demonstrated in this country should more or less allow people to make a good hunch how the racial polarization can be in the world's future. After the American Civil War, no racial conflict in this country should be termed as *war*, although bloodstain did speckle here and there from time to time. Indeed, even the Civil War cannot be remotely regarded as a racial war.

America started as a country that was established by whites. When the whites shed blood to gain independence from the United Kingdom, no nonwhites were heard to declare that they had been unfairly excluded in bloodshedding. However, when this country is handed down to newer generations with all the prosperity, whites are easily detected being hated for possessing so many goodies, politically and economically. Fairness is demanded from them, but fairness is not considered done unless the whites disadvantage themselves to some extent in all kinds of competition. For example, to get into the same university, they are required to either have higher points of merit than the others or let the others enter first with lower points; in political election, every race is allowed to use the slogan "Vote the candidate of your own race" but not white. The list can go on. The same mentality drives America to apply the same principle in the foreign affairs. All along, America helps to put a heavy pressure on the then white government in South Africa to relinquish the power to the blacks because the major population there is black. Looking back in her own country, however, white as a majority in population is not allowed to be a dominating race by nature. As this habituated idea about fairness between whites and nonwhites becomes more and more taken for grant, the whites are gradually put in a position to be seen as born losers by nature, on purpose or not. Thus, whites are more and more often disfavored in law or in public opinion worldwide, and particularly so in

America. In a 2009 ceremony celebrating the historical result of the American presidential election, some entertainment celebrity howled on the stage, "So good my president is a black." It was so natural for no one to ever mention it from the angle of racism. However, if a white entertainer ever shouted, "So good my president is a white," everyone could predict that he would lose his job.

America has been criticized as a country of racial discrimination, although laws after laws are seen enacted to suppress it. It is certainly true that she did have an infamous history of racial enslaving. However, the cold fact was that she also demonstrated to the world that she was the only country that ever pioneered the self-correction on political ugliness. The cost she put up and the extraordinary bravery and nobility were all found no comparison in the world's history. Very unfortunately, and it should be shameful to the entire mankind, for this or that kind of purpose of some people, only her less-brilliant part in the past has been found endlessly emphasized, either domestically or internationally. Gratitude and admiration to those who had their chest ripped open for the others' liberty are scarcely detectable, either in school textbooks or history documents or common political comments.

During the election year of 2008, a white Christian clergyman was revealed in TV preaching in a black community, advising his audience, "White people say, 'Do not hold me responsible for what my ancestors did . . . but you [the whites] are the beneficiaries of what they did!" This is a strong message from the black community to the whites. Here are some questions raised from this teaching:

1. Do we hear love and forgiveness or hatred from this message that is given in the hall for God?
2. Does this message mean that white people's possession has been in the watch list?
3. Does this message also mean that when conditions mature, all white men's possession can be considered as having been resulted from criminal activity?

After this was revealed, the response from the black community was that this phenomenon has been quite popular all over in the United States, with an attitude that could be read as "accept it or else, why be so surprised?" Racism, racism: people all thought that they were working hard to have you removed!

Judging from the appearance of this white clergyman, he should look no less than a well-educated man. He must know clearly the meaning of beneficiaries, and beneficiaries of what. A question here needs an advice from him: if the black slaves had not been beneficiaries of the bloody Civil War, would he rather have the Civil War not happen at all so that the black slaves could stay at where they had been? One may say that the white started the Civil War for their own economic agenda, but not initially aiming at liberating the slaves. This might be semitrue with two catches: First, before the war started, an outstanding slogan from the North was to abolish the slavery system in the South, not a slogan of economic demand. Second, that a person joins the police for his own interest should not be used to tarnish his heroic contribution in cracking down crimes, nor it should be used to devaluate the benefit a victim thus receives. One should clearly know that there was not even a slavery insurrection to have the Civil War triggered and started. Joining the police force or joining a robbery gang, both motivated by one's own interest, has a social value of day or night.

The whites launched the worldwide antislavery movement beginning the later part of the eighteen century, not even because of any unbearable political pressure from the slaves. However, modern education seems to be giving people impression that there were only white slave owners, but no white emancipators. American education seems to tell people generations after generation that the slavery shackles on blacks were not broken until 1963. No one says that the liberty has been complete because of the Civil War, but neither should any one deny that it was a historical turning point. Without this turning point, all later efforts of another one hundred years to perfect the liberty, by both whites and blacks, are impossible to be realized. For this or that kind of reasons, no document was found to present the blacks as a major component of man power in the Civil War, a war aiming at their liberation. This is not to devaluate the contribution that the blacks made toward their own liberation. However, this must make the following questions worth answering: If the whites were the major component of man power in the Civil War, why all their descendants must be held liable to what was done by a handful of whites few generations ago? How would one be seen as a beneficiary if his great-grandfather shed blood for other's liberation? How would one become no beneficiary after he received his liberation?

The beneficiary issue can go on. In 1860, one year before the start of the Civil War, in America there were about 31.4 million of whites and 4.4 million of blacks. If it is said that all the 31.4 million whites to be beneficiaries taking advantage of the blacks, can anyone submit a calculation to show how one person can successfully enable 10 people to have their living thriving? Furthermore, even if all the 4.4 million blacks had been enslaved, the number of white slave owners must be substantially smaller than this figure. Given a ratio of 20 to 1, that would be 200,000 slave owners. That is about 1 slave owner in every 150 whites. Why must all 150 whites be regarded as receiving benefit only because there was one slave owner among them? Why must all 150 whites be criminalized only because one criminal is found among them? If the whites must be seen as a criminal race because of a handful of them committed crime, should people be forced to accept a principle that any race that has occupied the major population in prisons is to be seen as a criminal race? During the Civil

War, standing on the side to liberate the black slaves were not limited to the whites who directly participated in the Union, but also the whites in Europe who supported the course of the Union. That the whites in Europe declined to accept the import of materials from the South was certainly a great help to the Union, thus a great help to the slaves.

Another concern about beneficiary could get answered from the growth of population of each race. In 2006, the population for American whites was 221 million, and blacks, 41 million. If the population ratio of white to black was about 10 to 1 in 1860, the ratio was declined to approximately 5.4 to 1 in 2006. How would one determine that a race has been made not a beneficiary in a certain society but a victim when its population is found rapidly increasing? Truly, in many measures, it can be said that the average whites have been enjoying a better lifestyle than the average blacks. The rapid change in population ratio between these two races must tell people that one of the cost for the whites to maintain a better lifestyle is to have imposed a genocide to their own race, although softly and voluntarily. Indeed, people expect the ratio continue to change toward less whites and more blacks.

In America, every one that meets a certain criteria of financial success must pay tax, and every one that meets a certain criteria of financial difficulty can get some welfare of this kind or that kind. If we average out the tax dollars paid by the whites and the welfare received by them per capita and do the same calculation among the blacks, which race would contribute more tax while burdening the society with less welfare claim? This question must be a taboo to answer, but it should offend no one to ask. He who declines this kind of investigation will only make himself susceptible when he claims he knows who the beneficiaries are. There is one way to avoid the taboo investigation, but a conclusion of beneficiaries can still be arrived at. Aren't the whites usually presented with an image of enjoying more because they had more unfair accessibility to wealth? Aren't the nonwhites, blacks in particular, regarded as ethnic groups that earn less because of the unfairness created by the society? With this reasoning, it must be illogical to conceive that the whites can be the ones to pay less tax but to have claimed more welfare while being assumed possessing excessive wealth. On the other hand, how is someone else not a beneficiary of the society while being able to pay less tax but then having more chance to qualify for welfare aid for his difficulty? Or do we need to straight out the ideas that whites are beneficiary because they are able to collect so much social aids without working and without paying tax?

Record showed that in history, the African slaves were mostly entrapped by their genetic "brothers" in Africa before they became merchants and transferred to a white dealer's hand. Truly, the whites intensified this activity by providing a lucrative market for this business to get prosperous. The bottom line question is this: why were these black genetic brothers left out to be counted as beneficiaries from the slavery practice? At the upheaval of slave peddling, there were three major worldwide routes for the current of slaves to flow out of Africa. However, historically only the one route going toward America and managed by whites was singled out for condemnation as if the other two had been so acceptable to everyone who never hesitates to show extraordinary justice in history. Is there a particularly reason to have only one culprit to be singled out for stone throwing? Can this be explained without an assumption of certain international effort behind this? Before even any reason can be given, a very obvious picture jumps into people's eyes: treasure hunters know who is rich but weak. A new batch of worldwide beneficiary of the slavery system is in the making, in the name of justice, at a time 150 years after the bloodiest war for slave emancipation is over.

Propaganda and education have customarily had people imagine that being a white in America would naturally enjoy more superiority and thus more advantage. But facts have a big distance from imagination. In a society as free as America, one can always encounter some genetic mix. So when someone carries a certain percentage of Caucasian genes is given a free chance to record his/her racial heritage, wouldn't it be only idiocy not to claim oneself as a white for the advantage? Surprisingly, most often than not, these mix youngsters would choose to identify themselves as a nonwhite, such as black, Indian, Polynesian, Asian, or at least, others in census, just not white. The existing law allows one as much freedom as one wants in so identifying unless a certain entitlement is involved, such as the Indian natural heritage, which requires proof of a certain percentage from the Indian gene pool. Do all these youngsters feel good to choose an identification that is conceived as leading to being discriminated? Please, do not insult their intelligence in making judgment.

Nowadays, we always find someone expressing, "I am a certain color, I am beautiful, and I am proud of my color." This kind of expression is allowed to be made by anyone except a white. If a white expresses being proud of his/her color, he/she will be labeled with racism. Why are the whites not allowed to feel proud while they have no less reason, if not allowed more, than any of the others to feel proud? The reason is because it does not fit into a scheme that would guilt-trip the Caucasians. Guilt-tripping is supposedly the easiest tool to bring down a proud head without a guillotine. It is amazing to see that the trick seems working very well. Many Caucasians do walk with a sinking chin because they are told that many slave owners did come out of their race in the past. Which race is absent of criminals, and should every member of every race feel guilty because of those criminals? The cool fact is that Caucasians do have ancestors who found that some among them had made mistake and were willing to pay heavy cost, including blood, to have the mistake corrected. Shouldn't the

Caucasians feel what they deserve more is pride in learning the courage shown by their ancestors? Is this kind of courage and nobility in self-correction seen in any other race in history? In contrast to this, using the shield (indeed a sword) of racial discrimination, some people nowadays do want to hide criminals within their group by demanding that police can only arrest criminal according to racial rate in population. Well, making the entire whites feel guilty is certainly serving to reduce resistance in cream skimming by someone else.

All aforementioned facts and advocacies against white only warn people of that a reversed amplitude shown by the racial pendulum has been displayed. Any discussion involving cultural evolution is seriously incomplete if the formulation of new racism can escape the social attention. This new racism has been taken for grant in many situations. Not only the whites must be "properly" trimmed in many standardized competitions, but they are also not even allowed to claim credit or to be seen as correct in some other situations. When a white presidential candidate mentioned with sympathy that blacks in many areas were not granted with full right to vote until a bill was signed by President Lyndon Johnson, this candidate was stormed with negative criticism. Why the criticism, isn't it a historical fact to mention? Did those who launched the overwhelming criticism want to tell people another history that all blacks had been granted the full right of voting before President Johnson signed the bill? It does not matter. If you connect a white with a positive value, you have a problem, and very likely, you are a racist. By the same token, if you do not automatically place a black in a position with more positive value than a White in the same competition, you also deserve a name tag no less than a racist. Indeed, some even claimed in the election year of 2008 that you were a racist if you were not to cast your ballot to a black candidate. Gradually, in people's concept, blacks should be allowed to be winners by nature in any game, checking any other merit but skin color is only a racist practice. If this is not racist, how about replacing the black with white with equal legitimacy in the above claim?

When a senator seat in Illinois was vacated by a black officer at the end of 2008, arguments like this shows up: "A seat left behind by a black should only be filled by black." If any white politician presents a similar argument for a white, his political career must immediately meet an end. Poll shows that 96 percent of ballots from the blacks were cast to a black presidential candidate in the election year of 2008. How should it not be seen as a result that has been racially oriented? In case 96 percent of white ballots had been devoted to a white candidate, resulting in a white taking over the presidency, will the election consequence be able to escape from accusation of racism, either at the national level or the international level?

It is not that the 96 percent figure reflects racism; it is that the same 96 percent figure on a different race potentially causing different social effect reflects racism.

—

Things are gradually getting more apparent: people complaining being racially discriminated are not playing fair game on racial problem. This further makes it clear that fairness pursuers that we mentioned few chapters back never intend to play any game fair. A feudalism practice of natural power inheritance is to be shammed into a democratic society. If the democratic society opens the door for power inheritance based on color today, what will prevent it from opening another door for power inheritance based on family name in the future? This is all what Socialism is about: starting from fairness demands, ending at automatic power grabbing. In this world, some country uses "democratic" to decorate the name of the country, but the third generation of the same family name is seen under way to get to the throne of absolute power. Is this the kind of democracy that some Americans are dreaming of? It is certainly worthwhile to notice that the intended color inheritance on power happens at the same time that racial proportion has pronouncedly shifted in America. Is the one way power inheritance model of South Africa also seeking immigration to America? It is also worth reminding that a racism practice that tolerates no criticism is far more dangerous than a racism practice that is unable to have all doors closed against criticism. Power inheritance by biological character will lead the way to the former.

It was still only in 1960s when a black must be escorted by military force in order to enter a university. Now, many whites are seen to feel the necessity to stay under the umbrella of blacks to reach for a certain favor in the society. Duke Lacrosse case in 2006 can serve a very typical example to illustrate this feeling in the white community. In this case, people related the DA's action with an approaching local election, and blacks were seen as an important source of ballots. For a while, a complaint was heard that the three boys' successfully escaping the charge owed to the fact that expensive lawyers had been hired by their rich families. Does this mean that someone could have successfully nailed these three white but innocent boys for the rest of their social life if they happened to be poor at the absence of evidence? Talk about racism, or antiracism in America! Nevertheless, rich or not, these three boys must owe their success in regaining freedom to the time in which they were born, but not in a time of fifty years later from today. Politician in this case shows the same pattern of political siding as shown by many American businessmen, who, in order to survive, frequently side with their opposite, the monopoly gangs of labor force. Is it for justice? Maybe. And justice is always found where interest is.

This is not the first time in history to hear criticism how wealth has helped the whites acquitting in situation similar to the three boys'. It has long been a tradition in many people's mind to explain the race ratio related to incarceration from the angle of wealth. It is absolutely true that it is racism if one particular race can constantly earn them more favor in court case because of their wealth, and such wealth must be associated with them. But how is it not also racism that skin color can be used to earn someone some favor in court case? One famous black athlete was *acquitted* of the murder of two whites after a *lengthy, highly publicized criminal trial* in 1995. However, a 1997 judgment against the same retired athlete for the wrongful deaths of the two whites was awarded in *civil court* by a jury. The athlete's success in the criminal trial left behind a question for someone to answer: is his wealth or his skin color the main factor in helping the successful acquitting? Wealth has been seen as a factor for whites to be generally assumed to have as powerful weapons in court cases. However, it did not show working for the whites in this criminal trial. Has the table turned? On the other hand, racism defense was one of the very moving factors for the jury in the criminal court to show this athlete the door leading to freedom.

Yes, infamous racism did exist in American history. However, the Duke Lacrosse case and case of the athlete's murder trial do also tell people that it has never gone but the tide is just washing in another way. Will the wave grow into a tsunami? Who can say with certainty that it will not? Not until the whites have a global organization of their own to promote the welfare of the Caucasians, the tide can only gain strength more and more intensely. The dreadful reality to them is that sources of pressure against them are not limited to one or two ethnic groups; the pressure is from a united racial front, which is invisible but with solid feelings from every corner. However, because of the potential defense that a Caucasian organization may provide for the frail white race, some people feel such an organization must be extinguished by every effort before it has a chance to start. The fate of the Caucasian club planned by the fifteen-year-old Californian girl aforementioned is a typical example.

Aren't people all too familiar with the term "white supremacy"? If all the aforementioned political behaviors displayed from the black community have had a chance to be demonstrated by the whites, no one will hesitate by more than one second to raid them with accusation of white supremacy. In addition to all this political behavior, black enjoys one more right that the white is forbidden to have: to organize according to biological background. Isn't it a naked supremacy, above all the political behaviors, intensified by governmental power? Compared to white supremacy, what supremacy should all these political exhibitions be termed under? If America has not been abducted by special interest groups, if America is not preparing to begin a socialist journey, if America can still declare herself a country with genuine democracy without blush, she should immediately introduce a precise and restricted definition of racism in her legal system. With this definition, racism in any direction, expressed with action or open advocacies, will be measured with the same standard by law—be punished with equal chance, equal efficiency, and equal severity. With this definition, all races truly enjoy the equal political right in every aspect. Any government office showing unequal attention in different direction of racial welfare is regarded as promoting and practicing racism.

c. Hatred

1. Birth of a New Vagabond

No one would argue that the drapery of modern civilization worldwide was pulled open by the people from European ancestral background. With the expansion of their civilization, white people, as a race, killed inside as well as outside their own race in history. However, in human history, which race or which ethnic group has exempted themselves from killing (or being killed)? Even today, as human beings all over the world scream loudly to haul the rule of jungle to a stop, we found Rwanda, and Sudan in Africa. Or how did the Cambodians lose one quarter of their population in only three years? Only a little more than seven hundred years ago, beginning from the leadership of Genghis Khan, Mongolians built an empire that occupied the widest land on earth in history. The cold-blood killing along their conquering path still left a nightmare, called yellow peril, to people in a big part of the world.

With all the killings that happened in human history, white's killing history is found particularly singled out and repetitively emphasized and hated in today's society. Indeed, comparatively, no one tries to disturb the nowadays Mongolian people with their past conquering history; and Genghis Khan is even viewed as a hero by many cultures. Why is the white treated so differently? There should be four reasons: (1) whites killed with high efficiency because of their advanced weapons in the colonization time; (2) after the killing, the landscape of the colonized land also changed abruptly with speed that is not seen before; (3) the white's highly developed civilization also makes their killing record with outstanding clarity in the books compared to the record made by other ethnic groups in history. However, all these three reasons are trivial in comparison to reason 4: the frail white lion still steps on a vault of abundant treasure, but his canines, though with numerous cavities, have not yet really felt off. To accelerate its succumbing, the best stone to throw at it is the clear record of their ancestors' killing made on to the others. During their colony expansion time, the whites' ancestors had left murdered victims in almost every race during the era of jungle, with some more than the other. It is only a matter of strategy of convenience to inspire all nonwhites to stand on the same front and single out this yet still wealthy target. The democracy advocated by the leaving lion even makes the attack seem almost riskless and costless. Of course, when there comes the time to split the treasure, the united front will also split. As usual in human history, hunters are then found to be too many, and the trophy is found too little, but that is in the remote future.

In humans' daily life, even without malicious wrongdoing such as killing, the status of being rich is enough to attract endless hatred. Jews in Europe are good examples in history.

Historically, in Europe there are two ethnic groups that found no land base to be called country of their own. One is the Gypsy; another one is the Jew, who does not have a country until 1948, yet it is not even in Europe. Both as political orphans on the land of Europe, each experienced quite different treatments from the locals. While both received intimidation including up to massacre in history, what the Gypsies faced was a pitiless attitude from someone toward a beggar, and what the Jews faced was a heinous attitude from someone who felt ready to tear a devil into pieces. Why is such a big difference? Frankly, no one is able to present evidence to show that Jews had ever committed racial massacre on any one to incur such hatred. The ultimate answer is, above all reasons including religious, the different quantity of wealth that each of them is able to command.

Jewish people, as an ethnic group, should be as good as any other ethnic group in Europe, if not said better. However, compared to the others, they went through a history with outstandingly more often calamities, which were not brought about by nature but by humans. They owe themselves being a magnet of these calamities to their ability, just like an elephant attracts poachers only because of its ivory. Their success in science may not necessary be a reason for them to procure all the troubles; the success in financial management surely was when they did not have a country of their own. A rich orphan must also be rich of greedy and vicious relatives. If the Jews ever had a country of their own, would they have to go through all these drastic events in history? In the absence of a mother land, they were constantly smeared, libeled, portrayed in pictures where ugly images were needed. All these libels were devised for one purpose: to legitimate an oncoming persecution that would lead to wealth seizing. What would the masses think about Jews if they could be successfully smeared as a gang spreading Black Death like what once happened in Europe? The consequence of the expulsion or even massacre following those smearing and libels was extremely sweet to the persecutors: an abundant wealth would be left behind by the persecuted.

If the Jews feel them having been mistreated by history, they will soon find another group of vagabonds walking in the path repeating their ancestors' history. Here a list is under fermentation, a list of replica of a history that Jews' ancestors wrote with tears and blood:

1. *Being a minority.* With the lifestyle that these vagabonds are addicted to, their females are unable to replenish new generations in number compatible to the others. To make it worse to themselves, they even have set up a system to encourage the rapid multiplication of the other races. They will soon found themselves drown in a gene pool of the others.
2. *No mother country to go.* With the thinning population, a democratic political system their ancestors pioneered must destine these vagabonds

to lose their government seats first, and countries second, to others someday. This, however, is still only an ideal picture to imagine. What if the others feel their population being large enough but unable to wait any longer? Cataclysm of anything must happen when a certain threshold number is reached.

3. *Facing insurmountable hatred.* If it is said that Jews are hated by many Europeans in the past history and by many of the Middle East people now, it can be said that these vagabonds are hated by many people all over the world. In order to make the hatred look legitimated, Jews were once accused of spreading Black Death, while a big sector of these vagabonds are very recently accused of inventing AIDS!

4. *Possessing wealth.* If these vagabonds had been poor chips, they would be able to find more unnoticeable corners to perish quietly like what Gypsies did. Jews failed to do this in Europe in the past; will these vagabonds be able to do this in the near future? Shrinking population but bearing conspicuously different skin color, shrinking habitats but carrying eye-catching amount of wealth, altogether would only make these vagabonds treasure islands of ease access in an ocean of envy. A few decades back, in some African countries, after the success of some "revolution," with only a few words like "this farm is ours, now you must leave," then a white must "peacefully" relinquish his ownership of a farm. One may say, "The land belongs to the indigenous by heritage." Good reason! Any reason backed by absolute power is good reason. In the future, when Europe is conquered in one way or the other, will the same reason work?

Is everything in this list familiar? They sure are. Jews, with their tears and blood, have prerun the history for these vagabonds in Europe. For these vagabonds, the lessons summed up in this list are free on the one hand but are also pricelessly expensive on the other hand. However, to any person, anything that can be obtained free is not worth being treasured. As to who exactly are these vagabonds, let us leave this A-guaranteed question for the second graders in the primary schools to work out.

What Jewish history shown in Europe in the past is certainly undesirable to the Jews. However, each time when torment descended upon them in history, someone may offer some help for them from somewhere in the world. The help may be far from being enough, but still better than nothing. In the future, will the same help come to the rescue of these vagabonds whose population would have been drown and the habitats would have been practically not there anymore? With the in depth permeation of races shown in the current world, the same thing can be wished to happen only if another virgin earth is available.

Indeed, when these vagabonds are in trouble, the future Israel may be put under critical test too. With the rapid shifting of racial proportion in the world, the country alliance available to Jews must also inevitably shift. Could people hold an optimistic view in predicting the Israel's future? The cold fact is that nowadays we see questions similar to this jumping around us more and more frequently: "Will France become another Algeria soon?"

2. Self-Decimation

Although the size of population means a lot of things in racial competition, whites seemed never hesitate to destroy their own people. War has been particularly frequent in the European history. Whatever the personality it is that has governed this race, such personality will sooner or later lead to the most devastating self-decimation in human history unless they are willing to unify.

In the twentieth century, there were three back to back events that practically brought themselves to their own knees on the population problem: the Bolshevik movement and the two world wars. Lenin and Stalin were no doubt heroes to the Communist believers, while Wilson, Churchill, and Roosevelt were no doubt heroes to the Democratic society. However, when taking Caucasians as an entire collective for consideration, will these men be held responsible more for the boosting or more for annihilating this race?

No one here said that the Russian tzar must be a good ruler. However, if the bloody so-called revolution did not happen in Russia, how was it necessary that more lives than the revolution devoured must perish? The lives involved with the Bolshevik movement can easily count to half a million out of a population of 3 million of Cossack population in the Russian Civil War, 20 million lives in the Holodomor in a Ukraine famine alone, saving hundreds of thousands of lives that were once Stalin's followers in his Great Purge. Do not forget that for each patch of "enemy" to be eliminated, there must be a compatible number of lives of those "comrades" to be sacrificed too. Needless to say, the very primitive motivation of Bolshevik movement was agitated by the hatred against the riches. Nobody could say that the rich should have some luxury life like what is found in those years; but neither can anyone deny that blood can only be more profusely gushing out when hatred dominating the removal of the unfairness. To make it worse, the end result is that the imagined removal of unfairness leads only to a more contrast unfairness.

If Adolph Hitler died in 1937, he would have died with no less than a national legend in the German history; but with eight more years in his life, he degraded himself quickly to a hysterical war criminal of World War II. However, if he was allowed to defeat the Russian Communists, who could say that the unsurpassed casualties on the war front in the West Europe would not have been saved? Who could say that the Russian casualties must exceed what had perished in history if the war was able to end on an earlier day with Germany appearing as a victor over those Russian Communists? "Winning over Russian Communists?" asked someone with skeptical eyes wide open. Yes, winning over Russian Communists! The war launched by Germany against the Russian Communists, if won, should have been the least life-costly opportunity for human beings to solve the problems brought about by Bolshevik. More than abundant history has proven that Bolshevik is no more than an organization who regards a life not being a life but a chip in a political casino for someone to ascent to dominance. Germany ended up as a loser because America and UK sided with the Russian Communists. An immediate gratitude made by Russia for their siding is to slide away half of Europe west to Russia, a quick follow up of Communist conversion in China, and gushing open two floodgates for blood to overflow in Korean and Southeast Asia. What a defunct mathematical result from the free world!

If it is said that the participation in WWII by the United States and UK was for defending democracy, the report card at the completion of their effort must say that they have defeated what they intend to defend in a big time. The very primitive motivation for Britain to launch a war against Germany was the crisis feeling of seeing Germany growing into a potential power to have the European continent united. It was even more intolerable in case Germany had a chance to dominate a land that Moscow was part of it. If it was not for this kind of motivation, it was hard to explain why British had chosen a much more expensive price tag for her future existence. The price tag from the Bolshevik group was to "liberate" the entire mankind; and such price tag had never promised to have Britain excluded in the list to be "liberated". On the other hand, the price tag from Germany was a bigger market share in which the Great Britain may well be a partner as well as a competitor under the same marketing principle. Together with the Germany price tag was the removal of Bolshevik; not only the removal was almost cost free to the Great Britain, but such removal would also clear the way for British businessmen to better exercise the marketing principle in the world market.

Almost the same fear and hatred found in WWII against a German dominance in the European continent hurled Britain into WWI. It was even sadder to learn that the first families in both countries had very close blood relationship. It may be a traditional national policy for British not to allow the appearance of a sole dominance in European continent. Each time such a potential appeared, Russia happened to be on their side of alliance. The same happened when Napoleon appeared to be a sole dictator on this land. Maybe history has jokingly designated a misfortune to Britain. She did have many chances to unify with France, but the more they struggle for it, the more English Channel seems to be widened.

Of course, a stronger Germany must mean a greater restrain on the freedom of Britain in doing business in the world. However, has the disastrous choice on the price tag during WWII gained her a bigger freedom? The answer is that the freedom is unfound; what is found is the disappearance of her empire. To make the choice more dramatic was that, as the war went on, a then-supposedly heroic leader asked Stalin for forgiveness for some acts twenty years ago. How was it not seen as begging in Stalin's eyes? If it was for defending democracy, a chief of proven ruthless killer gang would be the last one a democratic warrior would beg. Imagine that a father dislikes a young man, who, if not stopped, may become his son-in-law and later takes his daughter away. And then, this not so strong father begs a proven rapist to give close custody upon his daughter against the young man. Alas, let us talk about political genius!

If Germany were allowed to win over the Communists, there might even exist a slim possibility that Holocaust against Jews would not have come to such a large scale as recorded. The Holocaust served Hitler a big purpose: to extort for the urgently needed money to support the war, but Jews happened to be in such a convenient position for him to target at. The Jews had been historically slandered, lacking a mother country, and worse of all, commanding attractive amount of wealth. Contrast to what Hitler did, the second man in Hitler's regime, Hermann Goering, together with some other important government members once thought in another way. They considered that the enormous Jewish labor force was an asset too valuable to waste while Germany was preparing to invade the Soviet Union. It may be possible that if Germany achieved her goal on an earlier day in fighting against the Communists, Jews may have to sink into a long-lasting misfortune of slavery, but how was Holocaust not worse than this slavery misfortune? Or even, when Germany had a chance to reevaluate the contribution that Jews made toward her victory, who can absolutely exclude the possibility that the Jews would have been awarded to some extent? No one should say that Jews deserves any misfortune, but what could have been worse for their suffering than what was found in history?

Because of the large base of population in Asia and the strong agriculture tradition of desiring new offsprings there, population in Asia bounced back quickly after WWII. No such luck for the Caucasians in Europe. They can never recover in the world competition on population after the three major population disintegrating events. Their failure in such competition is further reinforced by their addiction on the consumption of cheap labor.

3. New Hatred

When the old hatred is found having finished its job, a new one is seen needed to be brewed by someone. Today, with a new mask, hatred continues to direct the world's affairs.

After WWII, the hatred from the Communist world against the capitalist was overwhelmingly intense. After the collapse of Soviet Union, such hatred has been subdued to a certain extent. But other forms of hatred immediately filled up the place thus yielded. The major currents of hatred that can be obviously found in the contemporary world include the following: Islamic hatred toward the West, together with the Jews, the blacks' hatred toward the whites in America, and the worldwide hatred toward America. Going along with all these currents, hatred against the Japan is also detected time to time, but the intensity is still no comparison to the other three. All these hatreds have one common characteristic: those to be hated happen to be financially more successful. Of course, those launched the hatred would say that there is legitimate reason behind all hatred.

Legitimate reason? Let's see.

So please someone give a reason, a reason from reality other than ideological teaching, that Jews must be hated so much as to be necessary to have them removed from the world. (Although the original wording is more like to have Israel rooted out from the map, but when it is given a chance to become true, what would follow to happen to the Jews?) Reason but only from ideological teaching is not legitimate enough; it is only a reason leading to an argument that "my God is better than you God," waiting for arbitrary determination, including violent force.

So good reasons were searched or invented to buttress the hatred toward the American whites. The hatred from the blacks toward the white has been planted so deeply in American history that Abraham Lincoln is singled out as the only hero ever to liberate the slaves as if all whites else had been slave owners. Yet this is not enough. Someone feels the need to invent more reasons to elevate the magnitude of the hatred. "AIDS is invented by the American white government to kill the blacks," shouted from a pulpit in a church from the black community. Isn't it a miss that obesity is not also included as a reason for the condemnation? At least, obesity is not found in Africa; if it was not for aiming at blacks in America, why was it only found in America?

So America is hated worldwide for her to plug the Communist leaks all over the world. How isn't she a warmonger by sending so many of her citizens to Korea and Vietnam to shed blood? How isn't she a criminal to have saved a power in South Korea that so many North Koreans risk their life to escape to? How isn't she a monster to have prevented so many countries in South Asia from sinking into the control of Red dictatorship? How isn't she a robber by paying so much money for obtaining petroleum in the world but others cannot (yet)? How isn't she an imperialist to have so many oversea military bases that are desired by local powers?

4. Who Needs Hatred?

Everyone can find something that can cause discomfort in his life. Such discomfort can be elevated to hatred with ideology that is uniquely possessed by human beings. However, to most people, they know hatred only to serve negative purpose in life. Knowing this, most people try to refrain themselves from allowing hatred to develop without bound. Many religions, typically Christianity, emphasize love, gratitude, and forgiveness in life. However, no matter how noble the teachings from various religions can be, their teachings are good only within the limit of personal conducts. When confronting with political behaviors, these teachings are usually overpowered. Opposite to love, gratitude, and forgiveness is hatred; therefore, overpowering them must also be hatred. So it is logical that political interest needs hatred. A typical political example of this can be illustrated with the outcome of American Civil War.

With a rare found political jewel—"Malice toward none . . . . Charity for all"—as a firm believer of Christianity, Abraham Lincoln finished his job as a victor in leading the nation to end the slavery system on this land. He even lost his life as so many slaves should have been celebrating their liberty. With so much sacrifice from those who contributed to the liberty of the others, hatred against the descendants of the liberators should have never have a chance to enter people's imagination. Conflict may be still surfacing up here and there for various reasons, but the profound benefit received by someone should have had enough power to soften the unsatisfactory emotion and let brotherhood atmosphere prevail in all kind of negotiations. However, in a history of nearly 150 years after the Civil War, Lincoln is found being the only hero singled out to be hailed. Thankful statement toward liberators is almost undetectable. It is unreasonable to believe that the liberators must continue the hatred upon those whom they sacrificed so much to liberate; it is equally unfair to say that those who received the benefit of liberation are ingratitude people. However, it is certainly surprisingly disappointing to see that hatred is rooted so deeply in the society. It is even more surprising and dreadful to see the hatred explode so unrestrictedly behind all kind of school textbooks and so many Christian altars. Why do things happen in this way? The answer is that politicians of special ambitions, either from white or black, need the hatred to stay and even to be amplified whenever possible. Any thankful statement can only hurt their agenda.

So even with a war that must be seen as justice by any measurement in human history, hatred cannot be removed but fertilized; can human beings expect not to have hatred produced from any other conflict? No. Ordinary people do not need sublimation of hatred, but they must live under politicians' dominance. Without hatred, too many politicians would have found themselves unemployed. The Bible or other religious doctrine books do teach people how to live without hatred, but the politicians found no words from these books regarding how not to plot with hatred. After the Independence War, George Washington did not ask people to extend their hatred against the Royal throne into another period of American history; neither did Abraham Lincoln ask to extend the hatred against the South enemy. But these kinds of politicians are so rare to find in history. We must notice that while not asking hatred to be extended, Washington was also a politician who declined continuation of power many times in his career. Contrastingly, while advocating relentless hatred against unfairness, all Socialists, without exception, are people who must seek for continuation of power, either to himself or to his family. Isn't it enough to tell who needs hatred and what it is for?

Frankly, preaching hatred enables a person to easily access a beneficiary position: reaping nobility of sympathizing misfortune victims, earning mask of justice, collecting feedback of contribution from victims, such as respect, followers, and, subsequently, power in various degrees. Given that most of the people feel unsatisfactory in life most of the time, to agitate and to echo resonance of hatred is a much easier job to do than the opposite: to preach love and gratitude and forgiveness without hypocrisy. This job requires demonstration of generosity and sacrifice. Because of the demanding nature of sacrifice, ordinary people, who are usually governed by human instincts, must truly have the gut, intelligence, wisdom, unselfishness and persistence to pick up such job and do it well. Now, are we surprise to see so many teachers in school choose to preach the liberal agenda, in which hatred enjoys much higher priority? The answer is that excessive "freedom" of speech allows them to choose a much easier job for the same pay. "Liberal agenda preaches love, too," someone protests, "People working for such agenda have shown unusual dedication to, for example, illegal immigrants, who have been entrapped in such miserable life." However, do they preach love without hypocrisy? Or do they preach love but at the expense of the others' sacrifice and then reap the benefit out of the support from the illegal immigrants for themselves? In comparison to this hypocritical love, we easily understand why Washington and Lincoln truly earned the reverence from their countrymen, for they knew when and how to put an iron fist on the country's enemies, but never asked to extend the hatred to the enemy. They know hatred destroyed love, which was what a healthy nation needed, and was also what they showed to the nation. Indeed, they did not even ask people to hate what was wrong but just to remove what was wrong. So, in comparison, we cannot read better personality than inexcusable selfishness out of those who preach hatred. Unfortunately, subconsciously moved by human instincts, so many people feel so easily to confer the wreath of laurel of nobility and justice to the wrong people, who know no better to satisfy their own selfishness at other people's expense.

Country is actually an animal territory marked by, instead of biological discharge, ideology and guarded by force up to the maximum possible violence. Before a country is finally settled, countless families, groups, tribes, ethnic communities on the same land must have been made eradicated. It would not be mistaken to assert that human beings' ancestors started killing competitors on day one that stone tool was invented. In other words, if human beings must nowadays clear all debts of ancestral hatred, they do not have a clear historical ledger to follow, a ledger that may span as far back as three million years ago. On the other hand, it may even be truer that each of the modern human beings is a descendant or a beneficiary of a possible ancient killer rather than a descendant of a victimized family going through massacres generation after generation. If someone must ask others to pay the debt of hatred in these few hundred years, how would his justice guide him to pay the debt back to those victims who were removed by the ancestors of this justice person many thousand, or million, years ago?

Chapter 10

Human Instinct, Racial Instinct, and Ideology

a. Instincts and Cultural Evolution

Like any animal, human beings have natural instincts. Fundamentally, there are three:

1. To extend life by all possible means
2. Dominance over the others
3. Desiring maximum enjoyment while keeping maximum distance from obligation

In addition to natural instinct, human beings have ideology, which is found absent in other animals. Humans' natural instincts constantly program ideology and ideology allocate humans' natural instincts. Because ideology is profoundly a mental factor, when it allocates instincts, it must be coded by the instinct to reflect a certain biological background. Different race has different biological background; racial instinct is thus resulted.

In the medieval time, the subject of geo center was not allowed to be investigated too much in depth and thus became a taboo. Similar to that, the relationship between gene and behavior is also a subject that has never shown too much tolerance for academic investigation. Because of such intolerance, investigation regarding how gene is related to the social behaviors shown in the following paragraphs must face resistance; and social behavior of each race is historically directed to look at from the point of view of environment, but not gene.

Being in the same environment of America, some races have higher percentage in prison than their norm of percentage in population as residents across the country. The same races, however, show up in universities with conspicuously lower percentage in population. If someone intends to investigate this contrasting phenomenon with academic tools related to genetic inheritance, he will be found politically sickened. In order to avoid being suspected as having been politically motivated, his direction of investigation must be politically guided. The most acceptable guideline is the so-called social engineering or, more directly, racial discrimination. How does discrimination work out? How would have such social engineering function globally in the entire human history up to three hundred or four hundred years ago? At the absence of such social engineering in such history, how did the landscape and real estate value of each continent end up with dramatic difference? With the explanation of social engineering, one must expect a race as a prevailing race in the society if such race can produce the most powerful politician for the country; one must also expect that races that could not produce athletes with the highest pay in the country must have been victimized, at least so in the athletic professions.

Too often, when people mention cultural evolution in the future, they focus on how technology and science knowledge will influence us in the future. Eyeing on the technology and science development, someone brought up an idea that humans will progressively have a larger brain in the future. Will this happen? No, because both human beings and Mother Nature have taken away the environment in which only the bigger brain have a better chance to survive the competition. Human beings have made all races highly permeated into each other in all habitats, and interracial marriage is becoming more popular. So, human beings have eliminated the environment in which brain size dominates the competition for surviving. As far as Mother Nature is concerned, she has placed human beings on top of a competition chain on which brain size and efficiency are the utmost determining factor. No other natural force will promote human beings to compete for a bigger brain. Similar to the nose size of those arboreal apes, it just stayed the same when the apes got into the wood where no natural force would force them to carry a higher-raised nose. With a brain size that was about three hundred cubic centimeters bigger than ours on the average, have Neanderthals left behind some offsprings showing even bigger brain?

When we mention cultural evolution, we need to be aware that human beings' cultural evolution marches at a much greater pace than their biological evolution. On the other hand, however, if we regard the entire collective of humans as being composed of a great variety of racial components, we must say that the biological evolution is also significant as the percentage of racial constituent gradually shifts. For example, imagine that one thousand years from today, all Caucasians phase out from earth. Archeologists at that time then find no concurrent human being possessing a skull that has an opening for air intake as similarly small as nowadays' Caucasians; shouldn't they conclude that human being's biological evolution has taken a dramatic change within only one thousand years? The racial shifting and the impact on cultural evolution accompanied with it seem lacking enough attention in all past academic discussion while such a shift is currently going on with ever-escalating acceleration.

Cultural evolution simply comes down to one principle: if there is no human, there is no culture. Therefore, the future outcome of cultural evolution must be determined by the numerical balance of future racial constituents. A very simple and indisputable observation is that, six hundred years ago, a population of almost 100 percent European Caucasian converted the European land with European style of landscape, and that a population of almost 100 percent blacks converted the African land south of Sahara with African style of landscape. Supposed this world is limited to only the lands of European continent and African continent and one of these two races must phase out on earth, what makes us to believe that the remaining one will not command the future cultural evolution with the same behavior and mentality with which their ancestors had converted their habitats? On the other hand, with the black population rapidly dominating, why should not they be allowed to bring America to a future with more resemblance to Africa? In fact, in a city like Detroit, America, where blacks have a dominant proportion in population, maintenance on buildings that have been incorporated with European traditions is found showing with less and less European emphasis. Therefore, it is not realistic to explore how science and technology will develop in the remote future without exploring the current racial ratio shifting. Simply, how do we predict with certainty that when a race governed by Islamic belief overwhelms in Europe, things listed in the next paragraph will not happen?

Churches in Europe must all follow what had happened to Hagia Sophia Church in Istanbul, Turkey. Art creation in any form devoted to non-Islamic belief will follow what happened to statues of Bamyan Buddha in Afghanistan. The presence of male doctors will be forbidden in birth caring while woman are not allowed to go to school to learn medical skills. Deep-space research cannot be carried on because television watching is prohibited. To make it even more dramatic, all Christians must convert to Muslims; voting in establishing government is necessarily discarded, as it can be viewed as a teaching not from Allah. All European languages that are involved with Latin origin are viewed as languages leading to some corruptive culture and are thus . . . Pork chops, bacon, and ham will permanently disappear from the dinner tables in a big area on earth; subsequently a big proportion of owners of pig farms all over the world must figure out what else to raise. DNA related to pig must become a taboo item in the laboratory, although some scientists think pig's blood may have a chance to help human blood transfusion in case human blood bank runs dry for any reason.

In case it is in a world other than Islam, what would happen if some powerful government agent considers research on AIDS medicine wasting too much money but tells people to cure the disease with sweet potato? Won't this give big impact in the future culture evolution?

—

After all these, do not forget the Mongoloids in the Far East Asia. Among this group of people, 1.3 billion of them once had a history showing extraordinary hostility to science. Scientists, engineers, professors, and teachers, except a few prominent and indispensable ones in military field, were all mowed from their career positions because of political need. Any pore from which knowledge can be licked was deadly sealed. Nowadays, although dramatic change has taken place among the people of the same geographical area, the governmental machinery still remains the same, and the same bygone leader still holds the absolute spiritual authority. Nothing can guarantee that their undeterred doctrine to "liberate" the entire mankind will not find a way of coming back; science is then found to be the biggest enemy of politics again.

Although all these patterns of cultural grouping seemingly happen as being independent from each other, don't we see that they are also highly geographically confined? This further means that all these are related more or less to some racial background or, even more frankly, racial instinct. When all of the above find their opportunity to happen, ideology would have restored its throne to dominate science and technology all over the world again, as in the long history before these few hundred years ago. Subsequently cultural development in the entire world will not continue to take the same course like what it is today, or to match what someone predicts for the future.

b. Intelligence and Ideology

Compared to any other living beings, human beings have acquired from Mother Nature one more but extremely unique talent to serve their welfare: intelligence. However, they have not acquired it free but with an extremely high price: ideology. In Mother Nature's selection sieve, before human beings appear, only the course of nature can exercise death, subsequently the outcome of fit or unfit, on living beings. However, ideology now shackles human beings with one more course leading to death. So selection by death (artificial sterilizing is too minor, ignore it) is no longer a job done by Mother Nature alone.

Both intelligence and ideology are mental products. Intelligence can lead to common sense, but ideology can lead to distortion of common sense, although unnecessarily so in all cases. For example, intelligence tells us $1 + 1 = 2$, but ideology from a tyranny can make people say they believe $0 + 0 = 2$ have more validity. The appearance of intelligence can be independent of human interest, but ideology can be found soaked with human interest. More often than not, however, it is difficult to distinguish between intelligence and ideology. The same woman can be said to be very beautiful by her birth child, but very ugly by her stepchild.

Whenever possible, ideology tries to overshadow intelligence. Possibly among all the mental products, mathematics is the subject that contains the purest degree of intelligence. Language, when stays at the level of word and grammar study and before it picks up the task of expressing idea can be considered very pure intelligence too. Medicine, physics, evolution, rhetoric—things begin getting contaminated with ideology. Art, music, literature are getting even heavier in contamination. Religions, which are ideas inseparable from political consideration, cannot exist without a skeleton of ideology. Then all political claims, discoveries, assertions, which are directly motivated and formulated by human interest and otherwise have no necessity to come to being, are sublime mental products of ideology. As such, it is not rare that some ideological products are profoundly buttressed with intelligence, but some are seriously devoid of intelligence. However, the more an ideology is found devoid of intelligence, the more it must rely on power for its continuous existence if it insists to exist because human beings cannot help but constantly examine ideology with intelligence. In this sense, the concept of so-called right surfaces up.

c. Obligation and Right

Mother Nature grants no "right" to any living being; she only gives chance. Without power support, no human being can have right, and no right is inalienable. He who has power support does have right that is inalienable to a certain degree until a stronger power overcomes him. No human can believe that he has any right that is inalienable when he gets into a lion's habitat while carrying with him only a copy of Declaration of Independence. Had America not won the War of Independence in 1776, all so-called rights mentioned in the Declaration of Independence would not have been so inalienable for anyone but the British king. By the same token, in case someday chance traps America into becoming a war-torn land, right that was once considered inalienable would become vulnerably alienable to everyone except few. With such character, the so-called right is actually a mental product embedding in ideology and is thus nearly devoid of intelligence.

Right is essentially enjoyment. When enjoyment is summarized as right in the human world, it must be restricted by many conditions, openly or implicitly. So right is another way to mean conditional enjoyment. Enjoyment requires the minimum spending of intelligence; and so does right within the boundary allowed by the restrictions. How would one apply intelligence when enjoying, or exercising the right of, sound sleeping? A person's most enjoyment-occupied period in his life is his childhood, and so this is the least intelligence-occupied period. Contrary to right is obligation. Obligation does demand intelligence for its fulfillment. While Mother Nature does not guarantee right to anyone, she does guarantee obligation to everyone. Right must come after the fulfillment of obligation but not before. The fulfillment of obligation does not guarantee the realization of right either. If someone can enjoy a certain right without fulfilling a certain obligation, it is only because some prerequisite obligation has been fulfilled by someone else. A cheetah that wants meat must chase hard after its prey, but nothing can guarantee that the prey will not be made lost right in its open eyes after the cheetah captures it. That a toddler cheetah can get free meat is only because the mother has worked hard for the baby. If a toddler cheetah whines for its "inalienable right" without the mother's work, death is most likely what it gets for its whining.

It can be easily imagined that a single human being has no right in the wild but enjoyment and obligation. The enjoyment is interrupted at where natural harassment appears, and even stopped if such harassment becomes irresistible. The obligation, on the other hand, never ends because Mother Nature never bestows anyone a natural environment that is free of harassment. When two or more human beings show up at distance reachable to each other, the enjoyment will be conditioned by each another's will. The concept of right thus surfaces up, regardless whether the right is verbally summarized or just silently indicated. Even who should have more chance to get access to a water source may be a big issue for dispute. The one who gets more often to the water source is not necessary a winner of right, as the issue may end up who will fetch water for whom.

Needless to say, grouped individuals are more powerful than any single one in resisting harassment and subsequently in pursuing better enjoyment. People grouping together for pursuing better enjoyment are then contracting each other, with contract terms both literally stipulated or silently implied. Subsequently, bound by such contract, government begins to form. Government's job is then supposed to mediate between social members regarding both obligation and right according to terms that have been literally stipulated or silently implied. Since the right is the result of power, such contract terms or laws are always drawn in favor to people who hold the power. Naturally, the more powerful somebody is, the more inalienable his right can become in the contract. Mother Nature has no chance to say a word on this matter.

That *"We hold these truths to be self-evident, that all men are created equal, that they are endowed by their Creator with certain unalienable Rights, that among these are Life, Liberty and the pursuit of Happiness"* can become universally applicable in any political system for those who hold power and be recognized as man. Therefore, the above quotation can also be interpreted as that the inalienability of each member's right is proportion to his rank in the social hierarchy. Different from all other societies, democratic society tries to broaden the definition of "man" to everyone in political terms, in terms of "government of the people, by the people, for the people." Therefore, democratic society tries to mediate certain right with the most rigid inalienability for the biggest proportion of her social members. By the same token, of course, if someone who holds power tries to mediate certain right in favor of a smaller proportion of populace in the democratic society, such person is making only a smaller portion of the populace to be recognized as man.

No matter how right is guaranteed, however, nothing can change the nature of right. Right is by no means any part that is bestowed by Mother Nature. It is one end of a social contract, which must be backed by power; the other end of the same contract is obligation, which is naturally assigned by Mother Nature and is further assigned by someone else's power, such as enslaving. Even at the very beginning of one's life, an infant's life cannot be seen as so much self-evident and inalienable without the protection from the parents. Besides needing power to back up, right has other natures: (1) it is not a material existence but a mental product; (2) its exercise requires less intelligence than the fulfillment of obligation; (3) right is something that no one would think he has excessively possessed (whether he would exercise the right is another matter). Compared to right, obligation is not thought of in the same way. Indeed, the power required for backing the right is by itself an obligation, an extremely essential obligation for right to be realized. Without it, no contract that guarantees right is honored.

In addition to all the four natures, right has a damaging nature to the society. Its disastrous nature lurks in the fact that the social contract in which right harbors can never be impeccably written with enough details to cover everything for everyone all the time. Simply, how can an infant write a contract? However, at the moment he begins breathing in the world, sadly, an infant from a slave family already contracts him as a slave to some slave owner. In other words, as soon as one enters a social group, he already contracts himself to the others according to a certain set of contracts; some terms of the contract are openly written, and some are merely implicitly indicated. However, as soon as a contract has space that is not specifically filled with words, it must leave room for ambiguity on both right and obligation for some people to take advantage. The damaging factor of right begins here: he who holds power always pressure for more additional rights as much as the ambiguity of the contract can tolerate.

In the so-called sweeping human right statement we just quoted from the Declaration of Independence, ambiguity on social contract makes it appear to have only the rights, but not obligations, be specified. Although obligation is not specified in this quotation, it is inconceivable for people to believe that they can materially "hold these truths to be self-evident" at the absence of fulfillment of certain obligation. On the matter of rights, "Life, Liberty and the pursuit of Happiness" are found only among certain unalienable rights in the statement. It means there should be more, but what are they? Again, ambiguity plays the role here. Is homosexual practice also among them? Is declining protection to homosexual activity with tax dollar or even with life on national defense also an inalienable right for heterosexual couples? Because human beings have idea, or ideology, social contracts of all kind must be under constant pressure to be modified, either peacefully or violently. When it is rewritten, it is basically rewritten in the direction that some group can then have more rights and less obligation. Otherwise, why bother rewriting?

Compared to other societies, democratic society, with government "of the people, by the people, for the people," must face most ideas to have the social contract rewritten with the highest frequencies, although seemingly most peacefully. Being pushed forward by the human instinct of "more right and less obligation," the repetitively rewritten social contract gradually contains more and more terms for "right" in more categories, covering more and more groups of people. When going extreme, right is a reflection of human greed; it is limitless. Subsequent to more right granting is more and more obligations to be sidelined. Excessive democracy thus begins.

The social contract is actually between citizens, ruling or ruled, but not between citizens and government. Because of the ambiguity of the social contract, in many cases, rights and obligation are only assumed to be there by a group of people but can also be assumed not there by another group. It is thus a battle of ideology. To determine who prevails, governmental power to mediate is needed, but a government can mediate only as much as her competence allows. When becoming incompetent, government assumes the role of an arbiter other than a mediator because of the abduction from some powerful group of citizens. This is what most is seen in countries ruled by tyranny. Even in democratic countries, with the same body of law and code unchanged, some powerful group can influence the government to function or not to function. A sentence "it is not America" from someone powerful can have the entire body of law written for centuries to protect the sovereignty of the country ignored. Too often, those who should have been facing the law for wrongdoing receive protection and profit instead of punishment only because of some powerful human right group's intervention. Selective application of law is a typical sickness of the government incompetence in the democratic society. Therefore, a murderer is able to escape death and even enjoys tax dollars from the victim's family to have his kidney replaced for a better life; it is only because some powerful one says capital punishment is "being inhuman." It is actually the traditional law for death punishment itself to have been put in death sentence.

"Ask not what your country can do for you; ask what you can do for your country." If it is said that during the John Kennedy's time, America was still fundamentally a country to emphasize obligation to have higher priority to right, the time has been gone. What flood in people's vision are banners with slogan of all kinds of right. Social contract, and subsequently ideology to match it, is at the verge of being rewritten in large scale. Right is seen more and more placed prior to obligation. Social contract is thus written upside down; an upside down society needs upside down ideologies. The upside down ideologies reinforces the upside down orientation of the society.

* When the life of a law enforcement agent is respected with less value than the life of a criminal, somebody is mediating the society upside down.
* When a victim's life can be stripped off in one split second and his right of life is kicked to a ditch but the murderer's desire of retaining his own life is fully granted by law, somebody is mediating the society upside down.
* When some group of people are so willing to stay "unlucky" forever while risking neither his time to work nor his money to invest, somebody is mediating the society upside down.
* When a liberator receives no gratitude and admiration but must be forced to groan with apology to those whom she liberated, the upside down ideology has confused the populace in a big time. With this upside down ideology flooding, people even see this is to happen: the blood-spilling liberators are requested to pay to someone who receives the benefit of being liberated.
* You know the society is upside down when marriage is seen only as a core of love between any two "lovers" but not seen as the only way to form basic unit for the time extension of a nation.
* When illegal border intruders are called patriots, how would the agents protecting the border be grouped: no less than treason?

Contrary to Kennedy's teaching, nowadays what people are practicing is to have the social contract rewritten as to read "ask not what you can do for your country; ask what your country can do for you."

* So newcomers to the country are not required to learn the host country's utmost primary language. Instead, the host country needs to dismantle the function of the primary language and let the newcomers' language overshadow it.
* So instead of warning a troublemaker of that a bigger rule would follow if he does not obey the small rule, the "right" believers send the troublemaker sympathy and sometimes even award.
* So everybody has a right to borrow money regardless his ability to repay until the national financial debacle appears.

Facts have overwhelmingly shown that democratic society allows terms for right in the social contract to expand without bound. Eventually, it must reach at some point that the binding force of the society can no longer hold. One of such events in America may be seen in this way: when tens of million "patriotic" border intruders (official figure is 12 million, popular belief is 30 million) receive amnesty, America would have accumulated enough political force to reconstruct an extremely fundamental ideology frame work: the Constitution. First of all, upon the amnesty becoming a reality, the fates of both Republican Party and Democratic Party are considered sealed but in different directions. Further, with a possible landslide vote from the new patriots favoring the party that fights for their amnesty, rewriting the presidential term like what happens in Venezuela is only a matter of time. Then, from that moment on, noble suckers: you have been whining endlessly for rights for too long; it is time to pull your thumbs out of your mouth and start to dig, either to dig for the foundation of your new king's palace or a grave for yourself, or both, but after the king's palace is finished.

It is said that American political system is a two-party system. Each party, either Republic and Democratic, has her own ideology and agenda. However, neither party can resist the abduction of ideology of granting more right. To survive in the election, right promising is a critical but also habitual political tactic. So each election campaign virtually becomes a rivalry of right promising between candidates. Gradually the ideology from each party must be required to expand and to include and to reflect right promising. So two more political terms implementing each party's ideology have become more and more popular in the political arena nowadays: "liberal" and "conservative."

There is no doubt and no need to explain that the so-called liberals are pioneers of the people who aim heavily on rights, various rights, and rights under all invented names at all cost (from others, though). And as customarily dwelled in people's reasoning, an opposite term must be there to contrast the liberal; and the so-called conservative is thus given to another group of people. Are the conservatives different from the liberals? Possibly. However, is the difference striking enough to enable them a flag leading the country to move in a different direction? Don't they dare! Here, for example, is a challenge for an American politician to pick up: which politician, no matter what sash he/she puts up in election campaign, Republic or Democratic, liberal or conservative, left-winger or right-winger, dares to mention to restore military draft for American defense at present time? He who shows positive intention on this matter before election must have a coffin nail put in his political career, a nail longer than that for either tax evasion or drug abuse or sex scandal. This is how damaging the right whining has been to the nation's mentality. Probably, to a big portion of the so-called conservatives in America nowadays, a title of *less liberal* would be a more accurate name tag.

Afghanistan and Iraq together have 63 million people. The Islamic extremists are no way the mainstream in these two countries, and thus the populations who are willing to work for them must be far less than this figure. However, motivated by a nonpopular belief, militants there can recruit undisrupted supplies of suicide bombers. In contrast, with 300 million populations, the American military in recent years constantly feels more and more pressure because of inadequate human supply to be recruited. It is said that some American politicians even begin to consider selling citizenship to foreigners in order to solve the problem. Alas! Does America still want to stay as one whole nation? Does she still want to own a military force whose soldiers truly feel that a wound inflicted on the nation is a wound inflicted on their hearts? Compared to the military forces of the East, the Western military seems to be traditionally having superstition on firepower. They pay far less attention on how human beings' psychological factor would influence the outcome in the battlefield. When America thinks herself as a superpower, and the entire world watches her as a superpower, but a war in Afghanistan must last for more than seven years and can only become worse, time to review the relationship between weapon and human factor is long overdue.

This is not to say that bravery is seen absent from American people. Indeed, mind you, no people in any other country are found to have combined their extraordinary bravery with courses of justice in so many events for such a long history. In the eighteenth century, they bled for their independence, setting up a brilliant democratic and free country in the world. In the nineteenth century, they bled for the liberation of slaves found on their land. In the twentieth century, they even bled for the freedom of people in other continents. On September 11, 2001, as the plot of some terrorists in UA Flight 93 was detected, roaring "let's roll," some heroic Americans immediately stood up and overpowered the terrorists in the high air, forcing the airplane to deviate away from a criminal route but to plunge in the wild. Their unconditional sacrifice must have had some extremely vital national buildings saved. In 2006 in Iraq, an American soldier named Michael Anthony Monsoor threw himself over a grenade to have its explosion smothered in a split-second decision, saving the lives of many others. The bravery they thus showed far surpassed that of any suicide bomber from the Islamic extremists in any measure. What more evidence should people ask for to judge the heroism and bravery of Americans?

However, different from animals, human beings all have ideology. Besides being pressured by surviving instinct to take part in a bloody fight, human beings' participation in a bloody fight is often a result of some ideological calculation.

Please tell an American citizen that the following will result to a good ideological calculation so that his/her life is well worth risking:

1. He may be put in a military unit that is in conflict with illegal "patriots." What is the antonym of "patriotism"? The consequence is dreadful!
2. He bleeds in the battlefield for the nation; the whiners of right may just release behind his back those captives in the name of justice. Of course, no distinct statement is necessary to tell him that what he did is injustice in the eyes of the whiners of right.
3. No torture suffered by American soldiers in enemy's hand receives as much national attention as the enemy's "suffering" in the camp that American soldiers guard. "Human right" constantly shows more regard toward enemy's soldiers with more value.
4. Joining the military is not something to be respected with honor but something so dirty that military uniforms must be seen to "get the hell out of our campus" from some universities.
5. Examples has been set to him that some forerunners had their chest ripped open for the liberation of the others, but it turns out that the liberated ones will not feel justice is done unless the liberators pay.
6. A country that the soldier guards with his/her life is felt intolerably sinful by a big population from this country, and this populace must swear "goddamn America" with their lungs exploding. So the soldier is either seen to have had defended a wrong country or the people he protects see him doing the wrong thing. To make it more horrendous to the soldier, some of the people he protects are doing two things at the same time: constantly getting a welfare check from the country while kicking butts toward the country like trash.
7. Won't a soldier feel being fooled by risking his blood in the battlefield only to find out that his loyalty to the nation is abused by the enjoyment of two groups of people: group 1 are those who practice irresponsible sex and flood the nation with fatherless children; group 2 are those who practice absolute barren sex with "pride," which only serves to discontinue the soldier's successors to come? What has the country left for him to enjoy after he drained his blood?
8. A soldier is defending a country, but the country has been constantly declining to defend even the language he feels dear ever since he learns to stare at his mother and listen to the mother's sweetheart calling. This language is not any other language but a language that has enabled his fathers of many generations back to build a prosperous and thriving country, a country that many people from other lands would risk any thing including life to join. No any other language has been able to

demonstrate more profound value than the one he has been taking pride. In history, so many nations would have asked their sons and daughters to defend the language when being attacked. However, at his generation, this poor soldier must stare right in his open eyes the fading away of the privilege of the language with which his mother reads him nursery rhyme every night at his bed during his childhood. Indeed, the country not only declines to defend it but even seemingly tries everything to dismantle it.

In short, a soldier's loyalty and blood are rewarded with heartless and pitiless mocking, insulting, or even penalizing. People trashing American traditional values for more right are granted with more chance toward the so-called American dream, but a young life devoted to the nation only get more chance to be trashed by those trashing the nation. The balance sheet of the ideological calculation is chilly to the young fellow with genuine patriotism.

It is true that, with America's current wealth, selling citizenship to foreigners may get Uncle Sam enough soldiers. However, to these imported soldiers, as soon as they complete filling the form that opens the door for them to earn U.S. citizenship, the transaction is considered done. The transaction is to enable them to live in this land but not to die for this land. Apparently, the West needs brighter mentality to survive.

Whether the nation is able to nurture more young men like Patrick Daniel Tillman or more young men like the young American Taliban fighting against his own country will make the outcome of the country completely different. One can assure that the more the whiners of right prevail in the society, the less the nation would find someone like Patrick Daniel Tillman to show up at the recruit station. When this continues, danger must someday arrest the whiners of right or their descendants (except gays) without mercy!

d. Life Span of Democracy

Human beings have accumulated too many schools of ideology. However, all schools only funnel all the ideologies to serve one purpose: how to distribute obligation and right among people. So far, it seems that human beings as an entire social body have traditionally emphasized fulfillment of obligation more than right, regardless who has received what. This only seems natural because this is the operating mechanism that Mother Nature has laid out for all her living beings to continue their existence. At the same time, however, this natural mechanism is seriously deformed in human society. The reason is that each human being has intelligence and subsequently ideology, which is programmed each other with human instinct of dominance. The general tendency of such programming is not only to enable oneself to enter a social contract with terms of the least obligation and the maximal right, but it is also exactly the opposite terms for the others. Exception does exist, such as that a mother tries to benefit a baby as mush as she can, but this is not general. Thus power struggle happens because everyone wants to have the most favorable terms in the contract. The concept of "unfairness" appears from the ones who are obliged to accept the less favorable terms. The problem is this: those who have been obliged or even overobliged are programmed by the same elements in the same way—to get the maximum enjoyment but to be obliged to the minimum. The concept of unfairness thus leads them to formulate another set of ideology: just to reverse what the powerful guy has advocated and forced upon, by all means.

So Thomas Jefferson et al were able to urge the maximum number of social members on American land to reject the obligation from the king by laying down a social contract that "we hold these truths to be self-evident, that all men are created equal." And Marx even tried to entice the maximum number of the social members in the world to enter a social contract of "workers in all land unite" with terms that offer "the proletarians have nothing to lose but their chains. They have a world to win." Covering people from different backgrounds, though, both types of social contracts emphasized right. This is exactly what unfairness removal is all about: to get more rights.

Can anyone imagine that a new social contract aiming at unfairness removal is to oblige people more? The problem is that a social contract emphasizing only promises of right without the backing of fulfillment of obligation must mathematically fail. Such contract will not match Mother Nature's operating mechanism. Therefore, without listing obligation fulfillment of the future society, the Communist manifesto, a social contract proposed by Karl Marx promising people "a world to win," must end up as a fake contract. The social contract that was proposed by Thomas Jefferson et al would have also been a mathematically fake contract if no obligation fulfillment is found. However, if people read the Declaration of Independence of America all the way to the end, they will find the brilliance of this social contract culminates with this contract term: "*We mutually pledge to each other our Lives, our Fortunes and our sacred Honor.*"

To Marx, the way for people to contract for future right is through dictatorship of proletariat; this concept is openly declared in the Communist Manifesto. In Jefferson's way, everyone pursuing the same right with the same inalienability must put up equal assets as collateral. In the case that is proclaimed by the Declaration, the most common, most fundamental, and most valuable items that everyone would have carried are "our Lives, our Fortunes and our sacred Honor."

Marx qualified the concept of dictatorship with legitimacy that was termed as being "of proletariat." Yet no matter under what name, dictatorship can only have the society repeat this model: the few who has more power to exercise dictatorship will have more right, and their right has more rigid inalienability; he who is allowed with less power must be obliged more. Therefore, Marx's way is merely to replace one type of unfairness with another type of unfairness, which is usually proven more vicious than the old one. "Dictatorship of proletariat" is a self-defeated concept. Dictatorship is made possible only after absolute power is established; absolute power is made possible only after absolute monopoly is achieved. Proletariat, on the other hand, by definition, is "the class of industrial wage earners who, possessing neither capital nor production means." How should a mathematical logic tolerate the ideological concept of everything being possessed by an entity owning nothing? This self-contradiction is therefore absolutely devoid of intelligence. It is a concept that allows the contemporary powerful few to exercise dictatorship over the maximal populace who are forced to fulfill all obligations for an "ideal" future that can only be found indefinitely remote. No concept that can be more deceptive in human history is found; yet no concept that has been believed with so much illusion by so big a populace in the world is found either. As we pointed out few paragraphs back, the more a concept that is found devoid of intelligence but must insist to exist, the more it needs power to back up. Since 1917, the nearly one century of Communist history with 100 million perished lives has offered nothing better to prove this point. Indeed, the higher fidelity a power shows toward this concept, the bloodier the power has shown.

Seemingly fair to a maximum of populace, Jefferson's Declaration of Independence is seen as having a serious pitfall too. What is the pitfall? Would Americans ask themselves: When they cite that "we hold these truths to be self-evident, that all men are created equal . . . Pursue of Happiness," how many times they would also have cited that "we mutually pledge to each other our Lives, our Fortunes and our sacred Honor"? The pitfall is not the birth defect of this document, but is so resulted by people's omission, on purpose or not.

Contemporary political life only finds that the obligation part—i.e., the "mutually pledge" part—is more and more forgotten, ignored, sidelined, deviated, violated, or even betrayed in American society. If the pledge is taken away, the support of the Declaration is thus taken away. A social contract without support is useless no matter what inalienable right it promises to protect or deliver. Only an idiot, but no one else, is to expect to get benefit from a contract without securing it by power. The power to secure what the Declaration promises to deliver to American people comes from nowhere but from their "mutual pledge." To abandon this pledge is to fatally castrate the Declaration. Therefore, the more we abandon this pledge, the closer we move toward a national suicidal ending. There is no other path we can walk while safeguarding the nation and subsequently ourselves.

There are ample examples to tell that the "mutual pledge" element emphasized in the Declaration has been more and more frequently and mercilessly trashed in every possible way in American political life.

Typically, example 1: trashing the mutual pledging of life. Murderers' wish to extend their life is constantly seen granted in court with various reason, procedure, or manner. This granting, regardless its motivation, obviously nullifies the significance of mutual pledge of life between a victim and a proven murderer. However, the mutual pledge stipulated by the Declaration has conferred right to no one, neither the judge nor the juror nor any person of "noble" spirit, to hold someone's pledge less accountable than the other's. To hold someone's pledge less accountable than the other's is to remove the law base of the society of genuine democracy. How is it different from a tyrannical society in which a more powerful person is held less accountable for his wrongdoing? Even worse, in democratic society, it is the murderer's life that receives more regard and thus more respect after the mutual pledge is broken. This is an open way to encourage crime; worse yet, this is an open way to destroy the Declaration. The essentially the same mentality found in such destroy has been so flooding in the democratic society that the idea to hold someone's pledge accountable in everything is found rapidly vanishing. The same mentality has been the major culprit that leads to the appearance of the recent financial debacle in America. We will come back to this matter later.

Example 2: trashing the mutual pledge on fortunes. When some citizens can develop a lifestyle by having welfare checks from the government without risking his time to work or risking his investment, the society has allowed them not to pledge their fortune while others were necessarily forced to. Truly, democratic society is no longer a democratic society if she walks away from a misfortunate citizen who is just struck by unexpected difficulty with irresistible magnitude, leaving him/her behind to perish without help. However, the help realized by giving a fishing pole or giving only a fish makes a big difference to the misfortunate citizen and to the society. By giving fish only, the society just makes the misfortunate ones more than happy to stay "misfortunate" forever. With the godsend misfortune, they even develop an abusive lifestyle, or worse, keep accusing the society being unfair to them with ever-escalating decibel.

Example 3: trashing the mutual pledge of sacred honor between citizens. When heterosexual group and homosexual group function so differently with their outcome for the extension of a nation, it is beyond comprehension to see that both these group of citizens have equally pledged their "sacred honor" to each other. It is better for each of these groups of people to make their own judgment with genuine pride on this matter: a nation dreaming of continuous thriving needs successors, supply and deliver!

Events similar to the above three examples are found spewing in the daily political life in the contemporary democratic societies. Some people, motivated by various reasons, noble or selfish, are certainly seen working hard to obscure as much as possible the mutual pledge from the Declaration of Independence. Without the mutual pledge, rights that are promised in this document would have been a political deficit. Without committing the mutual pledge, any right thus demanded is equivalent to a political extortion, ending up overcashing the freedom that this document is capable of providing. The ultimate end of this overcashing is to politically bankrupt this document and the nation. If this is not forcing the nation to commit suicide, what is it?

Compared to the time duration of domination shown by slavery system and feudalist system on earth, democratic system has not been lasting very long yet. However, she has been showing sign of being short life, appearing rapidly fading away from the global political platform. Although the England's Magna Carta of 1215 showed some democratic elements, it was basically a feudalist contract between the king and the lords. Possibly democratic ideals and practice were not really seen flourishing until John Locke (1632-1704) of England proposed the idea of three inalienable rights for human beings: life, liberty, and property. Only after three hundred years since then, we now must accept that democratic system, pioneered by the West, has been pressured by overwhelming hostility in too many fronts, externally and internally. If she continues being unable to produce another group of outstandingly brilliant politicians like Washington and Lincoln on time to save her, it is only a matter of time when she will collapse.

Democracy has proven highly energetic and benevolent to the masses. However, she has her Achilles' heel. Her Achilles' heel rests on the fact that her kindness constantly provides the possibility for her citizens to push her to a Utopian state. A big portion of people feel angry at her if such possibility cannot be made real. However, human instinct must destroy the Utopian state. The more a democratic society allows her to function like a Utopian state, the more incompetent she will become, and the faster her citizens, who must be governed by human instincts, will destroy her. The social element that she shows the most incompetence to deal with is the various conglomerated social powers, externally or internally. External incompetence can be seen through the various wars that America has been involved: Korean War, Vietnam War, War in Afghanistan, and War in Iraq. Each country opposite to America in these wars is seen much less potent, but so far, America can hardly claim as a decisive victor in each of these wars. Internal incompetence is even more eye-catching in people's daily life. The speedy conquer of American market by foreign merchandise and the outstandingly high number of incarceration in American prisons should need nothing else to witness her internal incompetence. The overwhelming flood of illegal immigrants on her land is an example of combination of her internal and external incompetence.

Depending on vote to get a job, each political candidate must make himself loveable to all conglomerated social powers, or monopolies, by promising them unlimited right, while vowing to put the obligation as distantly as possible from them. Again, right without fulfillment of corresponding obligation is a political deficit leading to bankrupt the society. If democracy is finally proven unsalvageable because of her Achilles' heel, her historical role can only be said to have genies of various monopoly to be unleashed, from capital monopoly then to monopoly of labor force then to monopoly of everything. The darkest time is yet in front of human society; the ancient slave time and the dark time of the medieval had nothing to compare. In the old days, when the weapons were limited to sword and shield, government may not have contrastingly more advantage over rebels in dealing with insurrection. In modern days, with the highly decimating power of weapons and the unlimited power of communication facilities, any tyranny can have absolute advantage over rebels in dealing with insurrection. Once a society gets controlled by absolute monopoly, the chance to get out is extremely slim. The collapse of Soviet Union is not because the people is invincible, as advocated by some eggheads, but is because the ruling class made some mistakes to themselves.

Slavery system and feudalism system are not welcome by the utmost population. However, they have been shown long lasting in history. The reason is that, being able to grab power at almost free will, the rulers dare to assign people with obligation while allowing very limited right to their subjects. As far as political resource is concerned, the political infrastructure is not overburdened, so the system can last. Sadly, possibly some of the naïve abusive right pursuers do not want to learn: even Socialist system may be predicted longer lasting than democratic system, unless some pitfalls in democratic society can be corrected.

Feudalist system in China lasted for more than two thousand years. Many families changed hands as rulers during this period. Although some families were seen lasting shorter than the others in ruling, the feudalist system stayed put until the destructive influence from the West arrived. Covering both slavery and feudalist eras, Rome also lasted for more than two thousand years. Ruling families were seen changing hands with higher frequency than those seen in the feudalist China, but the regime stayed almost the same. However, only in about one-tenth of the life span of the feudalist China and Rome, time already has made historians tempting to compare nowadays America and Rome. So how much longer will America last in people's mind? Seemingly, with the collapse of Soviet Union and her satellite countries in Eastern Europe, Socialist system appears to have met her end. "You wish," laughed the Socialists. Isn't

China, the most populated country, still there holding a socialist flag? After the ruling few in that country peek at the secret of the market principle of supply and demand and thus become more rational than Mao, the country has been lavishly evolving. She astonishes the world with new prosperity every day. While Soviet has disappeared in Russia, Socialist countries are seen bubbling up in South America. It means that Socialism has not proven failed, but the vortex just sweeps to some other spots of higher potential. All these happened at the same time that historians begin to compare America with the ending state of Rome. Why?

The opposite of democracy is dictatorship. Sadly, history has told us that democracy can joyfully give birth to dictatorship, but never the other way around unless violence is called for. In fact, when excessive democracy can force the acceptance out of a group of people who disagree on it, it is no longer that much of democracy anymore but dictatorship. What is dictatorship? The syndrome of dictatorship includes the following:

1. It needs governmental power to exercise.
2. As the exercise is in process, the will of the majority of people is ignored, but the will of a minority of people is emphasized.
3. The end result of the exercise benefits the minority at the expense of the welfare of the majority being damaged.
4. As the minority please, they can mobilize the governmental power to further enforce their agendas.
5. This is the most outstandingly critical symptom: dictatorship needs to twist concept. This is imperatively needed because dictators otherwise cannot apply political pressure in any degree up to persecution to silent the majority. This is the most direct manifestation of being devoid of intelligence about dictatorship.

Which of the above five symptoms is found absent in a tyranny society? It may be said that truth sometimes could be with the minority so that the first four symptoms are not necessarily the expression of a dictatorship. However, they must become part of the expression of a dictatorship when the fifth symptom steps in to play the role to pave the way for force.

In America, escorting non-Christian belief for prevalence over Christians by governmental power, in the name of freedom of religion, is a typical dictatorship developed in the democratic society. The oppression of the expression of Christianity is asked to be understood as the way to realize the freedom of religion; what more is needed to explain how concept has been twisted? We will have more on this topic later when we discuss the weakness of the U.S. Constitution. When illegal border intruders are called illegal immigrants, concept has been twisted somewhat; when they are even further equated with patriotism, people are asked to believe that moon and sun are the same things. All these twisted concepts need governmental power to enforce and the enforcing must in turn and unexceptionally benefit someone who is obviously not the majority. So is it still democracy that, pressured by governmental power, everyone must give up their common sense but accept someone's claim that a pebble is not a pebble but a piece of diamond? Is it still democracy that, pressured by governmental power, everyone must give up their training in math but accept someone's claim that $1 + 1 = 0$ is more valid than $1 + 1 = 2$? Is it still democracy that, pressured by governmental power, everyone must give up the definition from a dictionary but accept someone's claim that marriage is not limited to the joining of one man and one woman but the joining of any two lovers? If all these claims with twisted conception can succeed in securing acceptance from the majority of people, but such acceptance must vanish had governmental power not been there, why this kind of political practice is not dictatorship? When they succeed in all fronts, what is left for the majority of people to feel genuine democracy? Sorry, but silence! By that time, democracy has died. The same Constitution can still be there, but democracy would have only become "the emperor's new robe."

When the so called democracy is converted to be a social tool to guarantee the right that is insisted by some minority, but the same tool can be used by such minority to persecute any one from a majority, this tool is no longer genuine democracy but dictatorship. This is only a plain common sense. So relying governmental power for its agenda, excessive democracy is actually the same in nature as the traditional tyranny society. Of course, they do have difference between them, such as the following:

1. Traditional tyranny government has fewer people to enjoy the benefit of dictatorship, while the excessive democracy allows more (but still a minority in the society).
2. Traditional tyranny government dominates the society from top down, while the excessive democracy dominates the society from bottom up, or upside down.
3. Therefore, traditional tyranny government can demonstrate to the society with far more competence than a government that allows excessive democracy.
4. Therefore, traditional tyranny government may maintain good order in the society, but excessive democracy incessantly introduces more and more chaos to the society.

From the following scenario of property usage imposed separately by a tyranny government or a government supporting excessive democracy, people may recognize the difference or indifference between these two types of societies. In a tyranny government, a property owner may receive a note from the government: "You need to surrender your house to the government for a certain period of time with the minimum, or even zero compensation for the nation's welfare." In a society allowing excessive democracy, a property owner, without a note of forewarning, must surrender his property to a tenant who fails or even refuses to pay rent. To support the tenant's "right," the government sets up all kinds of legal barrier so that the property owner must further experience additional damage to have his property recover. So what is the difference to the property owner? The answer is the same damage but just through different agents. In a society like Singapore, which is considered by many as a government of dictatorship but with high rationality, someone who vandalizes a property is held accountable by a punishment of "cruel" whipping. In many cities of America, those who vandalized other's property are not held accountable (or even hailed as artists waiting to be discovered). Instead, it is the property owners who are held accountable by the government to remove the outcome of vandalism for the city to maintain a reasonable look. Isn't it obvious that the government has chosen to side with those who have vandalized to put under attack one of the three inalienable rights for human beings proposed by John Locke: Life, Liberty and Property? No more is needed to say how excessive democracy has dissolved the government's competence. A helpless government feels the need to damage its supporters to continue its ruling; how is it different from that a starving man feels the need to eat flesh from his own thigh to continue to stand up? If such incompetence expands to all fronts of the governmental operation, only anarchy is left for the society; what more is needed to issue a death certificate for this government? Anarchy is not democracy; excessive democracy is only extortion to democracy. The public who wants genuine democracy must stop all the excessive democracy before it floods the society beyond any control.

Chapter 11

Physics vs. Society

a. Concept of a Social Reservoir

In the world of physics, a state of higher energy, if not thermally isolated or mechanically restricted, would spontaneously evolve itself to a state of lower energy. The energy loss in the evolving process is manifested by either heat that egresses or mechanical work, or both. This is normal and universally happens with spontaneity and has been concluded by a set of thermodynamic laws. To return the physical state back to the original state of higher energy, the energy lost must be 100 percent compensated with additional energy to be spent to enable the returning.

Any human activity needs energy, and therefore it must be governed by such thermodynamic process. With human's intelligence, people usually classify energy as "useful" or "not useful." Useful energy is productive; otherwise, it is wasteful. Being productive or wasteful, however, is a human concept, an ideological concept. Mother Nature has nothing to do with it; she only keeps track how much has gone and how much is still left. Here is a simple application of this thermodynamic idea in human activity. In a community where modern mechanics is out of reach, if a man wants to eat the fruit of a walnut, he must lift a heavy rock then drop it to have the shell cracked open. When enjoying the fruit nut with the possible staring from the others, he is enjoying a right that he has created for himself; any process involved before the moment of enjoyment is a required fulfillment of obligation. So far, this obligation is required of him by Mother Nature. Furthermore, if it is not considered that the sun has stored some energy in the nut through some botanical process, the net energy he gets out of the entire process may be less than what he has paid in the obligation fulfillment. (How about he has been unlucky to have picked a fruit with big shell but rotten nut?) Indeed, the same concept is true even in the process of getting a small piece of diamond out of tons of rock or mud. If it is not human ideology that has assigned some value to diamond, such activity is a big energy wasting process. Will anyone crush tons of rock to get a walnut fruit?

Enjoyment of a walnut fruit is pleasant, but the pleasantness is so limited, and the process is so energy demanding and so tiresome. Isn't it ideal for someone that a satisfactory amount of walnut fruit is presented to his reach while not having to be involved in any shell-cracking process? So in the old time, some few people managed to force someone else to realize the process, and in modern time, machines are invented for the purpose. Activities in human society are far more complicated than we just mentioned. Even in the simple walnut fruit enjoyment, we have not yet mentioned the fulfillment of obligation of tree planting. The general idea is that enjoyment, or the so-called right in the human society, is an outcome extracted from fulfillment of obligation. From the point of view of the thermodynamics, right is the useful (a human concept) part of energy out of a total that must be spent in the fulfillment of obligation. In other words, energy available for right can never exceed the input energy that obligation requires. A society making right so abundantly available as to exceed what obligation can input is in violation of natural laws and must destruct itself until it completely stops functioning. Such stopping will be accelerated to appear if the society even tries her best to enable every social member to stay away from obligation fulfillment but to enjoy the most possible rights.

Human society can be regarded as an energy reservoir that has two openings; one is called the obligation input, and the other one is called right outlet. While the opening is wide open for obligation to be filled in, the opening of right outlet is depleting the reservoir. As explained above, for any amount of energy depleted, it takes extra energy to fill the reservoir back to the original level in addition to what has been lost through the right outlet. People who manage to stand next to the right outlet enjoy, i.e., deplete the reservoir; some others must keep spending energy to have the reservoir filled. Therefore, human instinct and ideology, under the grip of the law of energy conservation, naturally work together to group people at different openings with different chance. This is what the social contract is for: to group people at different openings with different priority.

We can further imagine the reservoir as a fresh water body. At the forever absence of rain, which can only be seen as energy of godsend, the only way to maintain this reservoir is to have someone constantly elevate water to it from a lower body of water. The water of the lower body is constantly soiled after the right enjoyment, and the water must be purified before poured into the reservoir again. If depletion is larger than input, there are two ways to easy the supply of the reservoir: one is to force the obligation fulfillers to work harder, and the other way is to shorten the distance between the reservoir and the lower water body. The first way must further intensify social tension; the second way is to lower the altitude of the reservoir, as raising the entire lower water body is more impossible. When the reservoir is finally lower to the same level as the lower water body, the reservoir to hold fresh water is gone; clean water and murky water can no longer be separated. The higher the proportional rate between right enjoyment and obligation fulfillment, the faster the reservoir needs to descent, if the society fails to entice the obligation fulfillers to work harder.

Democratic society is established on the idea to enable the maximal populace to enjoy the maximum possible rights but to be minimally obliged. If done in the genuineness that democracy is supposed to hold, the arrangement between these two groups of populace will be regulated by the principle of supply and demand as well as the principle of check and balance on power. However, human beings' instinct of the populace constantly pushes the society toward a Utopian asymptote. The consequence of this asymptote is to speed up the depletion of the social reservoir while minimizing the troop of the obligation fillers. When depletion surpasses the obligation input, the elevation of the entire social reservoir is forced to descend. There are only two methods to avoid the rapid descending: one way is to force the nature to operate as obligation fillers to maintain the input supply; the other way is to force people to regroup so as to reduce the proportional rate between the enjoyment and obligation. In some way, democratic society does do a good job in the first way because of her capability in liberating production through science and technology development. This also allows more people to leave the troop of fulfilling obligation but go to the group of depleting the reservoir.

As the troop depleting the reservoir expands, democracy brings people an illusion that everyone can be and should be limitlessly entitled to depleting the reservoir without obligation compensation. The obligation input and the depletion made by the right enjoyment are rapidly forced out of balance; even the productive operation of nature that is forced out by people's invention cannot catch up. People's greed can go over any boundary if not well controlled. The rapid excessive depletion of the social reservoir soon brings the democratic society to a state violating the natural law of energy conservation. Compared to democratic society, the other model of societies, such as slavery, feudalism, and even Socialism have potential to hold the reservoir longer, conforming better to the law of energy conservation.

After the social reservoir has descended to a level where outlet of right can no longer produce expected enjoyment, the society must spend huge energy to create a reservoir at a new elevation. During the process of reconstruction of a new social reservoir, people must be forced to regroup. Fierce, or even bloody, competition will show up. It is only a common sense that rebuilding a reservoir would be far more costly than retaining the reservoir at the original elevation. Therefore, the democratic society must constantly regulate, or willfully correct, the proportion between the troops of obligation fulfillers and the enjoyment depletion when peaceful opportunity is still there.

—

To say that the slavery, feudalist, and even Socialist societies may hold the social reservoir longer and thus have longer life than democratic society is not to say that these are better societies to the masses than democratic society. It does not matter. It does not have anything to do with better or worse, but has everything to do with how much conformity to the law of energy conservation each society would show. Better or worse, justice or injustice, love or hatred, noble or wicked, merciful or cruel, kind or devilish, they all are human values attached to a society; Mother Nature is careless about these concepts. What Mother Nature cares is whether or not any mechanism has been operating against her law of energy conservation. If it does, she would brake the operation pitilessly until it fully stops.

With the idea that the social reservoir cannot operate in a manner violating the law of energy conservation, we can compare two types of monopoly: monopoly of capital and monopoly of labor. He who has monopolized capital shows this price tag to the society: "Anyone wants enjoyment must provide satisfactory fulfillment of obligation." In his calculation, he must see to that the portion of reservoir under his control has bigger input than output. Although he may gradually increase the stagnancy to the society, he has acted in a way to retain more for the reservoir. On the other hand, those who have monopolized labor force show this price tag to the society: "Anyone wants obligation fulfilled must provide satisfactory enjoyment, or right." In their calculation for the transaction, depletion of the social reservoir has priority, they must see to that fulfillment of obligation will not exceed what they can deplete. Both types of monopoly follow the same principle in the calculation: "What comes into my control must be larger than what is let out from me." However, each type of calculation has utterly opposite effect upon the energy accumulation in the social reservoir. Mathematics must sooner or later present this question: how long can a society operate under the calculation dominated by monopoly of labor? Mother Nature has her hawk eyes on the energy bill.

In a nondemocratic society, authority places more people to fulfill obligation. They show to the society a price tag of the same nature as the price tag from those who monopolize capital. Obligation fulfillment always happens to be against human's instinct; it requires energy expenditure out of each fulfiller. This kind of will violating action slows down the circulation between the obligation and right and causes stagnancy to a certain degree in the reservoir. The more severely the authority is to violate the human instinct of the population, the more stagnancy would result. This further slows down the obligation input. However, the total number of individuals holding authority in the enjoyment position in a nondemocratic society is very limited. Quite often such a society is able to maintain obligation input larger than exhaustion called for by enjoyment. Hierarchy is a vital social tool in the nondemocratic society to make certain people to line up to fulfill enough obligations for someone to enjoy through the depletion, or to exercise right. The reason a democratic society is called democratic society is her diligent effort in eliminating social hierarchy as much as she could. The end result is more and more people lining up to deplete the reservoir while less and less people are found in the line of supplying obligation input.

In some way, slavery and feudalist societies are societies dominated by monopoly of capital, except that their monopoly scale are far smaller and more scattering than that is found in capitalist society. Unfortunately to capitalist society, she is the only society in which monopoly of labor is developed. So from the point of view of energy conservation, unless she can regain the control on the balance between obligation input and right output depletion, capitalist society has a tendency to shorten her life span through self-exhaustion. If remedy can ever be found to revive capitalist society, the remedy must rest on the disintegration of monopoly of labor.

Even Socialist society may have a longer life span than the capitalist society, although so many inhuman disasters have been found as a result of the practice of Socialism. Among all models of society, Socialist society is a society that has the largest and the most concentrated scale of monopoly on everything including capital. The single or/and few rulers in such society are virtually businessmen with guns. As businessmen, it is only natural for them to show to the society this price tag: "Anyone wants enjoyment must satisfactorily fulfill the obligation." However, as individuals having monopolized labor at the same time, such businessmen will not allow the appearance of the same price tag that is put up by the monopoly of labor in the capitalist society. Allowing such price tag obviously violates the Socialist businessmen's interest. Gun in hand, with everything under control, the Socialist rulers move the labor force in any way they want with the minimal compensation. The society shows the biggest stagnancy in human history. The consequence is that both the total obligation input and the total output of right are kept to the minimum. However, the overall obligation input is kept bigger than the output of right; violation of the law of energy conservation is thus avoided.

Democratic practice and low-educated community do not tolerate each other very well. The idea of democracy is to allow the maximal population to join the line in depleting the social reservoir. To satisfy this while avoiding failing, the society must be equipped with facilities to force the nature to have certain obligation fulfilled in place of human participation. This requires knowledge of precise operation and accurate prediction on courses of nature. Such knowledge cannot be achieved through cursing, spelling, sorcery, or bewitching, but only through scientific research and education. Lacking such knowledge, no one can force the nature to operate to meet his desire. However, greed for right to deplete the social reservoir is found the same in both advanced society and backward society; it is compelled by human instinct. Lacking the ability and proper facilities to force nature to fulfill enough obligation input, democracy in society of low education can only find obligation fulfillment being too slow but output of right depleting too quick. Democracy and a low-education community must then destroy each other. Sadly, this is exactly what the West, particularly America, is found doing: with undeterred diligence, she rapidly broadens her low educated or even illiterate social bottom by all means, including glorifying and promoting people of lower merits, eliminating competition in schools, limitlessly siphoning into her society with people of low or almost no education.

b. A System of Citizenship of Multiple Classes

In the nondemocratic society, hierarchy is built in a way that a very few people dominate over a large population and the ordinary masses are always over obliged. It is a society compelled by a penalizing mode upon the overall people. However, being benevolent to the masses, democratic society seems to meet an end quickly by deviating from the principle of energy conservation. So the hierarchy way with which the nondemocratic society is able to sustain its social reservoir in a positive state should be followed to a certain degree by the democratic society in order to overcome such deviation. Contrary to the penalizing mode found in the nondemocratic society, a hierarchy system in the democratic society should award and inspire the populace to fulfill obligation in addition to share the social reservoir. It then must provide no easy life for those who enjoy but avoid being obliged. Different privilege is bestowed to different class on the one hand; different restriction is applied to different class on the other hand. Standard of classifying citizens are basically related to personal conduct but not financial background or inheritance of any kind, making certain that the best class of the most prestige is covering most of the social members. This hierarchy system allows each citizen to know where he/she is and how to improve him/herself in contributing to the society. Do not feel strange about this system; capitalist system has long been practicing it except never openly and systematically stipulating it with law terms. We will come back to this topic later.

Hierarchy system according to citizenship classes is also what has been found practicing in China, indeed, in all Communist countries. Disregarding the conception of human value, this citizenship hierarchy practice did help Communist China to maintain her political competence and stability in three difficult periods that other power may find them impossible to endure. The first period is the civil war period; the second period is the man-made record national famine period from 1958 to 1962; the third period is the Cultural

Revolution, which lasted for about ten years from 1966 to 1976. What is this hierarchy system? The answer is the Chinese Communist Party!

All Communist countries openly declare that eliminating classes between human beings is their most honorable goal to aim at. However, these countries are found most sternly setting up classes with the most microdetails. At the very bottom of the society, members of the Communist Party are already citizens of higher class. As the social ladder moves up, the social space is more and more dominated by citizens who have earned the party membership, which endows a qualified citizen more privileges to rule and to enjoy. Even at the same higher tier of the ladder, a party member is far more powerful than a citizen who has not yet earned the party membership but political strategy happens to feel the need to place him in a certain office. A similar social ladder also exists among younger generation through the Communist Youth League, and among the children through Children Pioneering Team or something similar.

Openly, if everything is done honestly, a party member without official title enjoys no privilege over an ordinary nonparty member citizen. Joining the party is in a volunteering basis. Each of them gets the same pay for the same job like the ordinary citizen. During the past difficult times, they were not openly seen being able to have more substance enjoyment than the ordinaries. The difference begins here: as a member at the very bottom level, he is sometimes asked to show willingness to assume obligation, sometimes even very tough ones, without enjoying particular favor. In exchange, he earns respect from the nonparty member plebeians surrounding him; and subsequently a nonofficial leadership over a group of nonparty member plebeians around this party member is established. The unofficial ruling power is officially backed by the country's constitution, which stipulates that, taking China as an example, Chinese Communist Party (CCP) rules everything in the country. In other words, whereas no official title is given, a party member already has a higher political status than the other nonparty persons. So when a job is offered, with the concept that CCP leads (rules) everything, a party member naturally assumes the job. Indeed, sometimes a vacancy is even forced to leave vacant only because a party member is temporarily unfound, but nonparty members is doubtlessly unqualified, regardless.

In the old days, CCP monopolized every job; nowadays, some of the government jobs are open for nonparty member candidates if they pass certain examination. However, so many jobs are still beyond people's reach if they are not party members. Besides easier job access, a party member can share secret that shuts door to ordinary people. This door shutting thus provides tremendous advantage to the party members over nonparty citizens. With this privilege, the party members can have first choice on many types of merchandise, such as better house at a more favorable price, a better chance to know where a good job is available to their children. In the old days, the door shutting policy is an unannounced but officially guaranteed better opportunity for their children to go to universities. Nowadays, the door shutting policy is a guarantee of favor exchange between party members. However, all these are dwarfed compared to how much money or national resources some of them can command for personal purpose and how much power they can enjoy to settle argument in their behalf. As a Party member ascends the social ladder with so much "deeds" done behind door, he gradually enjoys a lot more rights than other citizens. Although gaining so much favor, they gain through ways behind the secret door. So his haloing impression of obligation assuming in the past days may remain untarnished for a long time among the masses. This haloing impression continues to benefit the legitimacy of leadership (ruling) of the CCP. Of course, earning a haloing impression for the CCP through obligation assuming is not always a requirement to join the CCP. For example, for a child of a certain officer, joining the CCP is only as easy as he is willing to ask. Joining the CCP is a direct way to join the ruling class. Through a system of awarding few people with leadership, actually ruling privilege, the CCP is able to continue to incite the people to provide obligation fulfillment with highly volunteering willingness.

Openly placing Communist Party members as top class citizens is only a "noble" and "legitimate" political operation in this kind of countries. Opposite to this, ordinary people are categorized into many classes, among which some of the classes are defined as enemy by nature and by virtue. A child born in the "enemy" family is a counterrevolution demon ever since the moment he separates from the mother's body. Besides classifying citizens, the Communists in dominance have a lot more heave-weight political equipments in hand. An efficient surveillance system, both openly and secretly, is one of them. Under this system, voluntarily or involuntarily, even husband and wife, parents and children can become spies on each other, politically betraying each other, let alone relatives, workmates, and friends, let alone the surveillance from the government to each citizen. Safety between party members is equally lost, not even the second man in the country can feel comfortably safe, because there is always someone above needing to know how each "screw" away from his vision is behaving and functioning. Another heavy-weight equipment is concept twisting, such as glorifying poverty at the social bottom. Glorifying poverty can be so forceful that everybody must feel the need to freeze a smile on his face to cover up the cyclone-like rumbling sound in his stomach, meanwhile chanting "my leader has made me the most fortunate guy in the world." One is thus made even to betray himself with the maximal cruelty. (But this is only one of the self betrayals the Communists need from their subjects, baby!) A concave "belly" showing the contour of spine must save its last breath for this chanting. If someone is found absent of this smile, the surveillance system is following to trigger the meat grounder to work. If one wants to commit suicide to avoid this society, the corpse he leaves behind will become a political skunk for his family. Every family member from then on will be listed as enemy of the country, as the suicide is seen as an indisputable evidence of hatred from this person and subsequently from his family against the leader, the Party, the "people," the socialist system, and the country . . . In order not to sidetrack our discussion, we are not going to dwell in this topic any longer; it can easily develop into another thick book. However, if the people in the West continue to allow the "fairness" and abusive right pursuers to increasingly pressure on their current governments, that type of government must seize them someday. Guarantee! Don't we see that the West has more and more glorified those people who show less merit, less willingness to contribute to the society? The Western world has been in some way approving Communism, matching the idea of glorifying poverty and descending the stair of concept twisting toward the hellish floor of the Communist club.

In comparison, in America, party membership of either Democratic or Republican does not bring the member any privilege over ordinary citizens. Because of this undifferentiated political attention over ordinary citizens, no people consider a party member being a pioneer on anything but a salesman of party agendas, or even just a political job applicant during election time. While the party system in Communist China can category people into different classes of citizens, the party system in America cannot. However, if grouping citizens into different classes of citizenship is found so effective in maintaining a positive social reservoir in the nondemocratic societies, why should the same be made absent in the democratic society? Otherwise, the more a society removes the social status differences between people, as having been shown in the democratic society, the more this society approaches Utopian, the more the society is paralyzed. As to how to set up such system, we will further discuss it in part III, "The American Path."

c. Also a "Golden Rule"

We have discussed few golden rules.

Golden rule number 1: If a society expects to continue, she must see to it that her reservoir has more obligation fulfillment input to it than right depletion exhausted from it.

Golden rule number 2: No reason is good reason unless it is backed by power, but any reason is a good reason if it is backed by absolute power. As far as this rule is concerned, the West has lost a substantial portion of power for self-sustenance. Time is found more and more to distance away from her.

Here comes the third rule, which may also be "golden": Giving a right to someone is easy; taking away a right from him must face resistance. With almost the same magnitude, excusing someone from an obligation is easy; to assign an obligation must face difficulty.

With all the erosion signs that the West has shown suffering, her future is critically linked to the success of overcoming this "golden rule." If she cannot overcome, the fate of the West is considered sealed. Together to be sealed are the fates of the Caucasians, belief of the Christians, and the so-called democracy and freedom.

The West starts her fate sealing by developing a culture that encourages uncontrollable greed; such greed is even reinforced by the sickening ingratitude to those who have offered obligation fulfillment. A nation that fails to control this mentality must fail herself. The reason is very simple; to encourage greed is to encourage excessive enjoyment. Based on the reasoning that enjoyment requires less intelligence, allowing excessive enjoyment to dominate is to allow dominance of intelligence of lower grade over the intelligence of higher grade. A government that is "of the people, by the people, for the people" is thus transformed to a government that is "of the children, by the children, for the children." A formula leading to a nation's being ruined is thus underway. On this principle of children ruling, America is seen competing far better than anyone else in the world. If nothing is done to curb it, it must sooner or later lead the way to a government "of the tyrant, by the tyrant, for the tyrant."

China, as everyone can see, has restored her status as a giant in the world and is on the way getting even more powerful. She has missed this status for too long. Last time when she held this status, she was a self-sustaining, self-replenishing, and self-locked country, exerting little influence to the world. This time, when she regains her giant status, she has her door fully swing open (economic sector only, though). Meanwhile, she has become far more rational than thirty or forty years ago too. If the world's political gravity must eventually shift to her because of her population and the wealth that the West is so willingly to lose to her, the world's tranquility will greatly link to her willingness to stay rational. Do not forget, she declares that she would never abandon Marxism. This must mean that she would never let go the goal to "liberate" the entire mankind. "The proletarians have nothing to lose but their chains. They have a world to win." (The Communist Manifestos) This Marx's claim is forever very tempting to the Communists. Their absolute dominance, the so-called dictatorship of proletariat, has been seen winning; what is wrong to win more? When they complete their "liberation," people will immediately find Marx's claim to be read as "the proletarians have *nothing to win* but their chains. They have a world to *lose*." That "they have a world to *lose*" has been partially worldwide proven by the Communists with two facts. The first fact is that all those Communist (or Socialist, the same) leaders emerging from guerrilla wars must seek to secure life time ruling power, or even secure the same for family hereditary. The second fact is that those who usurp power by taking advantage of ballot in democratic society must seek for extending period of personal ruling power whenever possible. When these leaders have won the world, what is left for the common people, the cheap labor providers, to win?

History has proven that the higher fidelity a government holds on the concept of dictatorship of proletariat, the more inhuman hardship people must suffer under this government. If it is not the existence of the West, particularly America, it is impossible to imagine China would have loosened the grip of the dictatorship of proletariat over her people. However, the dictatorship had really made her too dirt poor. "Being backward must be humiliated," said Deng Xiao-Ping in 1978, as this Communist leader launched the reformation in China. So more rational than Mao Tse-Tung, he felt the need to replace the straight jacket with a zoo cage over the people under his control. Some barbarian heavy weight political equipments, such as classifying people into enemy families, glorifying poverty, are put into history. However, in the future, if China feels strong enough and faces no more force to humiliate her, she will have full freedom to determine how her appetite of dominance to be better satisfied: to continue to be rational or to force everyone to the Communist heaven through the gate of dictatorship of proletariat. It is more likely that the heaven-and-gate option will satisfy them better because the other option would leave a zoo cage other than a straightjacket to the people. This is way less satisfactory to their addiction of absolute dominance.

There are abundant signs showing that the Communists in China make certain they have retained the full freedom to determine how rational they want to be in the future. Examples are: (1) Ownership of every single centimeter of land is tightly grasped in the hand of the government, and (2) her military force continues to be called Liberation Army, any suggestion of name change is abruptly declined. As to example 1, if any land a person stands above belongs to others, how much of this man's freedom must be linked to the mercy of the land owner? Nowadays, it seems that the government allows the ownership of private house. Do not forget, the land under this house is allowed to be occupied by the house owner for seventy years. At the end of the seventy year period, the house owner must either kneel for a new "negotiation" or move the house to Mars. As to example 2, Chinese Communist Party has "liberated" the Chinese continent for nearly sixty years, and people in Taiwan have obviously had a better living than the Chinese in the Communist continent. As far as domestic "liberation' is concerned, the army's task has either long completed or undesirable. If the Communists have not detected the need of "liberation" from beyond some domestic scope, retaining the name of "Liberation" for their military force is an obvious illiterate mislabeling. However, "liberation" is such

—

an attractive and legitimate word, its value is beyond description. In WWII, one of the "reasons" for Japanese warmongers to launch a war in Asia is to *liberate* the Asians from the whites' shackles.

Both of the aforementioned examples are too worthwhile for the world's people to watch out. History has shown that what all Communists actually love are the antonyms of the words that they openly advocate. When they say what is to be good for you, watch out, they are actually aiming at what is good for them. "Dictatorship of Proletariat" only ends up dictating the proletariats in all their countries, no exception can be found. Didn't recently we hear from them an undeterred chanting that, for the well beings of everyone on the globe, they would sternly oppose protectionism between all countries? All of a sudden, the Communists chant in the same common language with the capitalist countries after nearly a century of their uncompromised protectionism, which is a stone wall that even a migrating bird must hit and drop dead if happening to fly into it. Do we need to wonder why they suddenly show so must eagerness in concerning with the welfare of people even including those of the West? Answer can immediately surface up if one bothers to inspect how their wallet has been inflating and how the wallet of the West has deflated at the same rate. People can also immediately test their genuineness of sincerity in their antiprotectionism by asking them to remove their protectionism on their TV screen, radio, publishing, newspaper, Internet . . . Aah-haa! Isn't it clear? Only one hundred percent of protectionism about their means of expression can guarantee their freedom on determining how rational they want to be in the future. And so is true the other way: they already decide to retain the full freedom on determining how rational they want to be in the future, so they need to retain one hundred percent of protectionism on their means of expression now. The miserable extirpation gone through by Falun Gung, a loose and peaceful organization in China, has witnessed no better how much CCP wants to be rational, and whether being rational for a need or rational for a change in virtue. Permanent and absolute monopoly is the entire essence of Marxism. Anything that the Communists feel challenging to their absolute monopoly must be destructed and completely buried by all means; if not done today for a reason, do it tomorrow. To safeguard that purpose, Lenin declared long ago and nakedly: "All powers to the Soviet!"

The monopolization aroused by Marxism does not stop in the material world but must even extend to the ideological world. "Religion is the opiate of the people," claimed Karl Marx. Doing so, he did not expel what he proposed about Communism from what he claimed; then it became convenient for him to replace all religions ever existing in mankind with the faith he invented. What is religion? It is the set of faith by which a group of people believe to be empowered to have all human difficulties solved. As a religion, Marxism asked people to recognize their God on earth, but not in heaven like any others. To any other religion, the understanding about it can stay at the argument of being true or not true. To Marxism, its opening maw of dictatorship is disguised by a hologram of proletariat; until a person gets in, he may even decline to debate about its truth. If science develops to a stage that inserting an antenna into a human's brain can monitor one's inner thought and control one's will, the Communist dictators will never hesitate to apply this technology to all their subjects.

It is possible that, when Karl Marx proposed his ideas, he had good motivation at the discovery of some unfairness in the society. If so, however, his anger about the unfairness apparently impaired his ability in reasoning in the field of sociology; he saw it to be justified only if the society is to substitute the old unfairness with some more vicious and more horrendous unfairness. Such an "anger" explanation for the debut of his doctrine is derived on a less darkened assumption that he still possessed normal personality. If derived from a point of view that is more gloomy than anger, the explanation for the Marxism's debut is absolutely negative: a man with extraordinary reasoning power had single out an opportunity to ascent to the throne of absolute monopoly of power while many of his contemporary peers were still searching in dark.

The aforementioned negative explanation is not a baseless assertion. What good deed has Marxism presented to human societies ever since its debut besides 100 million perished lives and few absolute tyrannies? Marx's biggest supporter of his lifetime was Friedrich Engels, a bourgeoisie whom Marx's doctrine claimed as being the number one enemy to be removed. Engels was not only a son from a successful industrialist family, but he himself was also a big capitalist of a factory. If Marx's idea had any genuineness and sincerity in "liberating" the proletariats, why these two men of extreme opposite interest could come to join a perfect combination? If no lucrative benefit could be seen at the end product of Marxism, why Engels was so willingly to place himself, as well as his family's fortune, at a position to be abolished by Marx? This combination has a very similar nature with the contemporary phenomena in America and in the West: big capitalists and labor monopoly groups invest in the same politicians in realizing their interest. "The history of all hitherto existing society is the history of class struggles" (the Communist Manifesto). Marx and Engels were both right on that. What they did not tell people was that, regardless how one may detail in classifying social members, there were permanently only two classes in the society: the one that ruled and the one that was ruled. Human instinct of dominance must determine that these two classes could never be eliminated from human society. Hiding this fact from people, Marx and Engels actually also hid from people that the existence of the dictatorship of proletariats needed the existence of a class to be ruled, and that he who led the proletariats naturally dominated the dictatorship. Would these two men decline becoming "leaders" of the proletariats if the

Communist movement ever succeeded during their lifetime? Particularly Engels, he must either rule on top of the dictatorship of proletariat or to be removed at the success of Communist movement; he had no other choice.

Wicked and deceptive as Marxism and Leninism show in social development, however, the modern fairness and excessive right pursuers, intentionally or unintentionally, never cease to funnel the society toward what these two doctrines, actually one, advocate. Human instincts continuously disarm people's alertness against these pied pipers; only intelligence of higher grade can rescue them from being further sunk by their instincts. If the intelligence of higher grade can eventually be proven not powerful enough to fight against human greed, human beings must ultimately and permanently follow those pied pipers of fairness to enter a society of absolute monopolization. Indeed, even without the diligent work of the modern pursuers of fairness, human beings have long been putting up a history marching in the direction of concentrating power. How many countries were found five thousand years ago on the globe? Not too many prominent ones were found, but countless tribes were scattering all over. Now, after a history of numerous fierce competitions and negotiation of alliance, they all consolidate into about 190 power groups of equal dignity and legitimacy on the same surface. This number must tend to be fewer in the future.

d. Contemporary Social Reservoirs

Among all the 190 power groups on earth, there are three particularly large systems of ideology influencing them: Christianity, Islam, and Socialism, which can be often interchangeably called Communism. Among these three systems in the contemporary societies, Christianity is found to be most entwined and permeated with the idea of democracy in social structure.

So far, statistically, governments that are established with high degree of permeation between Christianity and democratic social structure are the most benevolent ones to the people. However, these governments are also the ones showing more and more incapability in retaining the social reservoir compared to the governments established with the other two ideologies. The personal conduct of forgiveness from Christianity is always interpreted into freedom of damaging or even destroying. The idea of holding culprit accountable is distorted as being inhuman; democracy is not genuine unless the government's ability in retaining and rejuvenating the social reservoir is allowed to be attacked or even dismantled. Therefore, the social reservoir of the democratic society is always exposed to full-scale destruction, from the spiritual world to the material world. Comparatively, the Islamic and the Socialist governments are far less energetic in rejuvenating the society, but they can function much better in retaining the social reservoir; therefore, they can both be potentially stay longer in the human evolution. Even under the same Christianity ideology, in the older days, before a country "perfected" herself with democracy, the same country retained the social reservoir with better chance, and the regime (not ruling family) stayed with more stability. Examples of how quickly democracy defeats herself in comparison to other social models are as follows:

1. Rome never introduced democracy to her society, but she lasted more than two thousand years. From the year that Christianity began to dominate to the year the East Rome ended, it lasted more than one thousand years.

2. Democracy in England began to exert strong influence to the Royal Throne beginning 1649 through the parliament. A partially retained monarch system and the prevailing dominance of Christianity did allow democracy to greatly rejuvenate the social reservoir. However, the further "upgrading" of democracy in the twentieth century rapidly mowed away her empire. A kingdom, counted from the year Rome left, lasted for more than one thousand years is seen to be fading away in less than four hundred years as democracy barges in.

3. Democracy ended the feudalism on the mainland of China in 1911; it ended itself on the same land in 1949 after only thirty-eight years of freedom to the people, allowing a more vicious feudalism to be restored with the combination of serfdom and slavery, termed as the "people's democratic dictatorship," later and genuinely termed as "dictatorship of proletariats." However, being more efficiently in retaining the social reservoir, such a dictatorship has lasted for sixty years now and continues to last.

4. Being able to be more rejuvenating, democracy entwined with Christianity once exerted heavy political pressure to the society that was dominantly Islamic countries. However, the rapid self-exhaustion of the social reservoir of the democratic society quickly allowed the Islamic world turned the table. Although being less rejuvenating, Koran teaching restricts greed from being wide spread in an overall scale in the society; it also emphasizes the idea of "hold the culprit accountable." It thus effectively retains the social reservoir. While the democratic reservoir shows more and more deficit, the Islamic reservoir manages to stay positive. After a suffering of the disappearance of the Ottoman Empire, the Islamic world is seen springing right back with full scale political pressure that the combination of democracy and Christianity feels losing resistance.

If the combination of democratic society and Christianity continue unable to curb the social epidemic of greed that they have wrongfully inspired, although unintentionally, and decline to enforce the idea of holding culprit accountable in all kinds of vandalism, damage, or destruction, the entire combination will die. With all signs that have been shown in the society, the ending of such combination can even be numbered by days. Too often, people from democratic society attribute the economic failure of Socialism to their smothering the market principle of supply and demand. But sadly, the democratic society is ignorant about the facts that the slaughtering on the same supply and demand principle has been noisily in progress in her own society. To make it even sadder, the slaughtering is not limited in the economic sector, but full scale in all sectors of the entire society. While the maximum populace is indulged with greedy demand, a democratic society, once stepping into the excessive mode, also tries her best to allow people to contribute the least amount of supply to the society. Through various social programs, such a supply restriction even extends to cutting off the supply of successors that the extension of a nation must need. The democratic society has been gearing up with full speed to shred the market principle and aim hard to drain the social reservoir.

Suicide has been pitilessly waiting ahead for the democratic society, unless something can be done and done on time.

Chapter 12

A Few Worthwhile Stories

It can be certain that too many readers have been familiar with all of the following stories except the sixth one. However, when we review them at a different time epoch, it seems each of them still reveals something that we may have neglected in our daily busy life.

Story 1

A scholar was approached by a desperate wolf. Tail between the hind legs, the wolf asked: "Can you save me, sir? Some peasants are chasing after me and want to kill me." Looking at the miserable wolf, the scholar was occupied by sympathy. "Get into my suitcase, hurry." The scholar's suitcase, usually for books, happened to be empty.

When the peasants arrived, they asked if the scholar saw the wolf. "It rushed in that direction and long gone," answered the scholar. Seeing the peasants disappearing, the scholar opened the suitcase and told the wolf, "You can come out, it is safe now."

"You are so nice, sir," said the wolf and grinned with his teeth fully displaying, "you must also feel glad if you could stop my hunger by becoming my meal."

Story 2

A scorpion wanted to cross a river. A mouse happened to appear and wanted to do the same thing. Knowing the scorpion would sting, the mouse tried to keep a distance from him. "Why are you afraid of me, brother, but you are so much bigger than I am?" asked the scorpion. A little nervous, the mouse just silently watched him. "If you let me ride on you and get to the other side, I will let you know where you can find a lot of food."

"You will sting me," said the mouse.

"No, I won't!" replied the scorpion.

"Will you promise?"

"Yes, I promise!"

So the mouse let the scorpion ride on the back of his neck and swam. When they were still in water, the mouse was thinking of the food that the scorpion promised, and the scorpion was thinking of his own safety, clasping tightly onto the mouse. Upon landing, the mouse asked the scorpion where the food was, the scorpion stung the mouse.

"You promised not to sting!" screamed the mouse.

"Sorry, this is my habit, and I cannot change," said the scorpion. After the scorpion made certain that the mouse's convulsion began to calm down, he got off the mouse's neck and left.

Readers, can you combine these two stories with the following attitude change?

Twenty years ago, "Please let me in, I will work for you, serve you. I will do any job you people don't want to do."

Currently, "We came here not for our culture being tolerated. We came here for our culture being accepted."

Could it be that twenty years from today, "We came here for dominance?"

Story 3

"Captain, a worm is found in the bottom plank of the ship. I need to find a prick to get it out"

"Let it go. Don't worry. The ship is so big, how can it eat all the wood? Besides, a little worm has a right to his life too."

All of a sudden, one day, as the worm gnawed away more wood, the rotten plank could no longer hold against the outside pressure asserted by the water but snapped and blasted away. The planks that were immediately adjacent to the rotten one were pried flying by the water gushing in. As the hole got even bigger, more planks failed. Can anyone expect the captain to be able to stop the ship from sinking? Possibly, but how much more effort had to be paid than just pricking a worm at the beginning?

Story 4

In the Bible, there is a story about two women fighting over the right to one child. One of them was the birth mother; the other one was making a false claim as a birth mother. Lacking evidence to determine who should have the right to the child, King Solomon said he would chop the kid into two pieces so that each of them could have half of the child. This decision scared the birth mother to death. Almost fainted, she said she would give up the right and keep the child alive. King Solomon thus had a sound judgment to determine who the birth mother was, and the birth mother naturally got the child back.

A similar version was found in ancient China about nine hundred years ago. Instead of threatening to have the child chopped, a high officer judging the case put the child in a circle that was chalk-marked on the ground. The two women then were ordered to simultaneously pull the child out of the circle as hard as they could; whoever pulled the child out first won the child. Just a few struggling, one of them let go off the child. Asked why, the woman said crying, "Your Honor, watching the child crying so painfully, my heart felt torn. I could not go on anymore." The high officer knew who the real mother should be; then let her take the child home.

If you think nobility shown by love prevails in these two cases, you have not hit the point. The true morale is that nobility in politics has no value if it is not backed by power. If these two birth mothers, so noble to their child, were not backed by some powerful authority, they must have lost their child. Nobility in politics without power backing leads to the same consequence that idiocy would lead to.

Story 5

About nine hundred years ago, the then emperor Gaozong of Song Dynasty settled a new capital in southern China after half of the land and two precedent emperors were captured by enemy. Would the catastrophic loss revive the court's will aiming to restore what had been lost, or at least to strengthen what was left? Not a bit. The life in southern China was too abundantly comfortable; sense satisfaction came before anything else. Indeed, generals keeping the heat of war against enemy were seen too troublesome and some of them were even framed and executed. In another 150 years, the final 11-year-old emperor was forced to commit suicide by plunging to the sea. Gone with him was the Song Dynasty, of course. The court had not been absent of officials who were able to predict what must be following. But why was anyone with a reasonable ability of thinking needed to be reminded of what might come after his family if his absolute power was removed by a hostile force? Gaozong and the subsequent emperors must be able to foresee a downfallen future, but the suffering was up to the future one to go through; why should the present enjoyment be interrupted by someone who is so remotely concerned in the future? But 150 years came by quickly.

Story 6

This author once saw a peaceful pig-butchering scene in some countryside. After feeding the pig a little, a man gave the pig a gentle massage. Together with the massage, the man also whispered at the pig's ear something so soft that even this man's wife may have never had a chance to hear. Dozing with its eyes half open, this pig, usually not too big in this area, was suddenly lifted by another two men and laid on a bench that was not too high from the ground. Before the pig could even respond, a brilliant sharp knife slid through its throat at no time. The man who swiftly sunk the knife was exactly the one who just gave the massage, for he had the free hand; while the other two men kept holding the pig in place. Just after a few strokes of its legs, the pig had its blood well filling a wide wooden basin that was quickly slid underneath when it was lifted. It was said that if a pig struggled hard before it was butchered, its meat would show smell of urine, and even more so would its blood. Not only the pork, but also the pig blood, was considered good food in this rural area. Why spoil either when only the least effort was needed to be spent? Possibly it was because of the way how a pig's mouth grew, when the final drop of blood drained, a trace of smile was detected at the corner of its mouth. How amazing! Oh no, but how noble!

*   *   *

In all preceding discussion about human evolution, we have hardly discussed how natural events would have impacted the evolution of human beings in the future. Given that natural events are essentially color "blind" in most of the cases (except, for example, sun ray reacting differently with various color of skin), given that human races are getting more and more infiltrated to each other's community, given that human beings have been place themselves on top of any competition chain among living beings, natural events can find it harder and harder to pinpoint any particular race to exert its influence. This is to say that, if any particular race has more chance to be threatened or favored in the future, it can only be the consequence of some cultural events other than natural events. After any particular race is eliminated or favorably multiplied, the overall physical feature of the entire mankind must be substantially and correspondingly modified. Mother Nature seems more and more letting go of her job in further "carving" humans, but lets human beings "do it yourself." She has more and more inclined to seat back and focus mainly on the energy bill for the entire mankind. However, she must have some cards we cannot predict; we all are living beings in her barn.

Part III

The American Path

Chapter 13

The Nation's Birth Certificate,
A complete Social Contract

And for the support of this Declaration, with a firm reliance on the protection of divine Providence, we mutually pledge to each other our Lives, our Fortunes and our sacred Honor.

Following a long dash line, the above sentence is given a uniquely special space to conclude the entire Declaration of Independence; and preceding the signatures of the Founding Fathers, this sentence completes what the Founding Fathers expect this Nation to be, and how this Nation should behave in the future. It is an indisputable fact that United States Declaration of Independence is the birth certificate of this nation. As a birth certificate of the Nation, it thus has absolute supremacy over any subsequent documents, and all subsequent documents have no base to exist without this birth certificate and thus must be submissive to this birth certificate in all aspects.

* With the birth certificated so written, so concluded, so signed, the Founding Fathers give a holy name to this nation: the United States (of America);
* With the birth certificated so written, so concluded, so signed, the Founding Fathers announce to the entire world: This nation is a country established on the faith of Christianity;
* With the birth certificate so written, so concluded, so signed, the Founding Fathers oblige the Nation to use English and English only whenever and wherever understanding is needed between the Nation and her people who need the love and protection from this nation.
* With the birth certificate so written, so concluded, so signed, the Founding Fathers lay a discipline to all citizens that their inalienable rights are not free. To retain any of them, each citizen must support the Declaration and thus must *"mutually pledge to each other our Lives, our Fortunes and our sacred Honor."*

Any attempt that is found deviating from these principles and expectations must be only found from those motivations that are associated with hostility against the Birth Certificate of this nation. This nation then must consider such attempt being in a position threatening her existence; this nation has full right to examine such attempt and correct any action leading to the realization of this attempt by all means.

Now, let us read the most essential text bodies of the Declaration of Independence of the United States of America:

1. We hold these truths to be self-evident, that all men are created equal, that they are endowed by their Creator with certain unalienable Rights, that among these are Life, Liberty and the pursuit of happiness.
2. And for the support of this Declaration, with a firm reliance on the protection of divine Providence, we mutually pledge to each other our Lives, our Fortunes and our sacred Honor.
3. Signatures from 56 Founding Fathers, who all are Christians and Christians only.

The entirety of this national birth certificate rests on the fact that it is a complete social contract. It provides right to her citizens; it also demands them to have obligation fulfilled. If either one had been missed, this certificate must be self-defeated in serving as a social contract but becoming a fake contract, a bondage enforced by some tyranny. Being familiar with the rights mentioned in the certificate, however, how many Americans have been aware of the obligation fulfillment required by the same birth certificate: *"We mutually pledge to each other our Lives, our fortunes and our sacred Honor"*? Of course, a complete contract must be endorsed by signatures. As a complete social contract, the Declaration of Independence does have fifty-six signatures, whose owners are all from the belief of Christianity.

Chapter 14

A Nation of Christendom

A sole being, represented by the word "Creator," as well as "divine Providence," is found in the text of the nation's birth certificate. This is a sole being that the Founding Fathers believe to have a nature superior to all human beings. Endorsing such certificate, all the fifty-six Founding Fathers are Christians and Christians only. Therefore, allowing no room for dispute, only God in the belief of Christianity can dominate the throne as a divine Providence, under which these Founding Fathers rallied. Therefore only God in the belief of Christianity is entrusted and relied on for the protection of this nation. If anyone must put up an argument that any name dominating in other belief should share the same throne with God in the belief of Christianity as Creator and divine Providence in the Declaration, he must first prove that the belief held by the Founding Fathers allows two or more Creators to be identified. Doing so, he is to declare that the Founding Fathers are apostates or at least infidels to their own belief. Can anyone do this? At the time this certificate was signed, without the signature of anyone from Islam, Buddhism, Sikhs, any name of any being, including Mohammad, Buddha, gurus, must be found absent in the Founding Fathers' hearts to guide them to do what they did.

Truly, the name of God in the belief of Christianity is not directly referred to in the document, but why should it be so needed? Why should a signature from a Christian be regarded as endorsement of other belief? Can a signature put up by a Muslim be regarded as evidence that he has placed Jesus or Buddha as his creator? When someone mentions his grandfather without explaining the existing nature of the grandfather, people will assume his grandfather being a human being in his mind, but not a crow, not a serpent; what is so unnatural? If someone shows his family name as XYZ without mentioning his ancestral origin, people just traditionally assume the family name of his great grandfather in the straight paternal line also to be XYZ; what is so unnatural? If someone refers to his own father without his father's direct presence, but someone else assumes such a father figure to be a baboon; besides insulting, what is it?

191

Very unfortunately to the Founding Fathers, after they handed down such a brilliant country to the current generations, if they were still alive, they could only find that "one Nation under God" is found intolerable by some of their descendents in American public schools. Apparently, someone prepares to deprive the Founding Fathers of their belief regarding how this nation is protected. In other fronts, someone is heard to propose to remove "help me God" or "God bless America" from governmental statement. If this nation believes that infringement of freedom of religion is unconstitutional, someone must explain why any of the following should be tolerated by the Constitution: the belief that the Founding Fathers held is deprived from them or is distorted as a belief held by infidels; the country they created is not allowed to keep in the same nature to match what their confidence once rested upon. In short, why is it not violating Constitution if the Founding Fathers' will is pitilessly attacked?

Should an attempt to change the will of the Founding Fathers be placed on a track to see green light? If the U.S. Constitution makes this green light possible, it must have failed its own function in protecting the freedom of religion, at least the Founding Fathers' religion. Nothing is crueler than that the Founding Fathers' generosity on freedom of religion is contorted as a reason and freedom for other beliefs to expel the Founding Fathers' belief. As far as the Founding Fathers' belief was concerned, they viewed the world to have only two religions: Christianity, which was what they held, and non-Christianity, which was what they did not hold but allow to stay without restriction. Just because the Founding Fathers allowed freedom of belief on this land, it does not mean that they allowed any one to deprive them of what they believed. Just because a generous home owner allows the free access to his refrigerator by a visitor, it does not mean that a visitor's hunger has first priority to be satisfied in his house. With the birth certificate of the nation so stipulated, Christianity is the indisputable home owner of this land, and all other beliefs are visitors and latecomers. When conflict arises between Christianity and non-Christianity, the birth certificate must request that the conflict is to be smoothed in manners that match the idea found in the teaching of Christianity. If any of the Founding Fathers had ever appeared in the ruin of the Twin Tower right after the calamity of September 11, would his "firm reliance on the protection of divine Providence" make him utter "Jihad"? Or if he had uttered "help me God" and gestured a sign of cross, must someone fire the Founding Fathers from the Declaration of Independence because of the U.S. Constitution sees the firing as "proper"?

Even after more than two hundred and thirty years from the day the nation was born, nobody is empowered to disrespect the wills of the owners of all the haloing signatures that endorsed the nation's birth certificate. The only way to disrespect them is to have the national birth certificate destroyed and have the nation removed. Imagine that the Founding Fathers have not yet died but age must

—

confine them in bed. "Son," a voice is heard in expressing a dying wish, "can you let me taste some candy?" "Sure," replies the son. After a candy is unwrapped, the son puts the candy in his own mouth and says, "Lick the wrapping paper, folk. It has the taste of the candy." Can we imagine a bigger insult than what this heir can do to his old man from whom he is in the way to inherit an abundant wealth?

> *Done in Convention by the Unanimous Consent of the States present the Seventeenth Day of September in the Year of our Lord one thousand seven hundred and Eighty seven and of the Independence of the United States of America the Twelfth. In Witness whereof We have hereunto subscribed our Names.*

The above clause is found as the last sentence in the original text of the Constitution.

If the word "Creator" or "divine Providence" in the national birth certificate happens to leave room for someone mistakenly to have illusion of ambiguity, the counting of years shown in the Constitution is exclusively governed by *our Lord* in the belief of Christianity. If, for example, the year counting was governed by the Lord from Islam, the year number must be about 579 less than what the Founding Fathers have accepted. Therefore, each of the signers must find himself humble and submissive to no one but the Supreme Being *our Lord* in the belief of Christianity. Indeed, the relationship so worded between *"our Lord"* and the signers in the U.S. Constitution tells people that any signer, if happening to come from other belief, must hereby make himself submissive to *our Lord* in the belief of Christianity when signing.

To reaffirm the spiritual foundation that this country is established on the belief of Christianity, the Founding Fathers even anchored in this foundation with the word "posterity" in the preamble of the Constitution. So worded, the Constitution is ordained and established to secure the blessings of liberty to the Founding Fathers and their *posterity*. By using the word "posterity," but not just offsprings or descendants, the Founding Fathers regarded themselves as being all from one ancestor. Naturally, by the belief they all held, the Creator in the belief of Christianity must be exclusively their only supreme ancestor. To introduce another image to share the throne occupied by Creator in their mind is to condemn the Founding Fathers as infidels to their belief. When the Founding Fathers used the term "our posterity," they even presumed that the future generations on this land were not only children of theirs but also from the very same ancestor. Therefore, every citizen on this land is first defined as a Christian by law, but then allowed to choose another belief without special procedure or ceremony if he/she desires to do so. The Founding Fathers have given the Christians not a single reason to feel powerless and paralyzed, but all the reasons and legal tools to maintain, or regain if ever infringed, their strength and pride on this land. As the most pious posterity, the Christians must feel that they have been obliged by the Founding Fathers to firmly unite, working together with all possible strength they can find from all fronts in the society, devoting themselves sternly for one common goal: let this country continue to be called

The United States of America

For the next ten thousand years, Christians are obliged not to allow history of Rome to repeat on this land; too many forces have attempted to scramble the above name to read "the untied states of America."

So in this nation, each citizen's unalienable right, if he deserves to retain it, is endowed by his *Creator*; this nation must have firm reliance on the protection of *divine Providence*; this nation's time is seen bestowed and perpetually governed by *our Lord*. All these beliefs have been inscribed as laws on this land and handed down to the *posterity* of the Christian Founding Fathers. Their wills are so recorded in the most sacred and fundamental documents of the nation with signatures that are soaked with the belief of Christianity. So taught and backed, if nowadays Christians continue to feel feeble and paralyzed on this land, they should examine if their minds have been compelled to accept values from other thoughts. So taught and backed, Christians should find no reason to yield to non-Christians on this land. So taught and backed, Christians must live up to what God and the Founding Fathers expect them to do. They must continue to dominate the spiritual world on this land and live up to what God and the Founding Fathers expect them to have: pride, a genuine pride that only an uncompromising guardian deserves.

"One Nation under God" must prevail again in America; "In God we Trust" must prevail again in America. They are so declared by the nation's birth certificate; they are so reaffirmed by the U.S. Constitution.

However, no one said it is an easy job. If a son feels he must come back to the house of his heritage and found the house being partially occupied by many others, what should he do? It is not easy for him to reclaim his most perfect ownership when confronting with the scornful look from all dwellers, who are there only because of the generosity of his passing away father. But it is only stupid and lame for him not to reclaim only because there is difficulty. Truly, upon reclaiming the ownership, it is unkind and impractical for him intending to stop sharing with the other dwellers. However, his priority to explore in every corner of the house must be fully respected before anyone else. At place he decides a cross to be properly placed, no belief can stop him. To stop him from doing this is equivalent to stopping the Founding Fathers from fusing the Christianity's belief with their signatures on the document.

Chapter 15

A Nation of English—and English—Only

The United States of America announced her existence in the world with only one language: English. It was so lawfully and imperatively witnessed by the Declaration of Independence. It was by means of English—and no other language—that this nation was given life by her Founding Fathers. The privilege and status as an official language for this nation was thus seized and monopolized right at birth by the language of English and no other language. The supremacy of English in this nation is unshakable, needs no dispute, and tolerates no seat next to her throne.

How is a language not a family language for a baby if this is the only language that enables the bonding of love and understanding between him/her and the parents at the very moment he/she takes the first breath in the world? How is the same language not a family language with which the parents set the rule and tell a child how to act and what he/she can have in this family? As the nation grows, the status of the sole official language for English in this nation is further reinforced by another two most authoritative documents in this nation's history. In time sequence, the first one is the Articles of Confederation, and the second one is the United States Constitution. When English—and English only—is entrusted as the indisputable linguistic tool to record all the three most authoritative documents to bring this nation into existence and guarantee her growth, how English is not the official language of this nation? Why more is needed to stipulate what official language should be for this nation? Indeed, by the same token, if these three documents were ever drawn with any other language, such as French and French only, and if someone now wants to declare English as the official language but not French in this nation, sorry, it's too late either by time or by law or by common sense. Indeed, among the Founding Fathers, plenty of them were talented linguists. However, among all the languages that they knew, among all the languages that had been existing on this land, only English is entrusted to establish the three most authoritative documents; all other languages were rejected. They did not even duplicate the same documents with any other language. Nothing is clearer: in their mind, such duplication is never necessary; and as law documents, such duplication is not even allowed. By doing so, they had kneaded English as an inseparable and irreplaceable part into the nation that is called the United States of America, but any other language is indisputably expelled to have a chance to become an element of the nation. To allow any language to steal the throne that English has dominated, someone must show some document written in that particular language that is even close in weight to the three documents in American history. Any such attempt must fail. The historical fact is that at the time the three documents were so drawn in English, the political status of any other language displayed no difference from nonexistence on this land. What bigger insult can this generation do to the Founding Fathers by dignifying something that they decided to discard?

There are two groups of people who are ignorant about the monopolization status of English as an official language on this land. The first group is those who lie to the face of everyone that English is not an official language of this nation. The second group is those who demand law stipulation to establish the status of official language for English. By demanding law stipulation, however, the second group of people gives an unintentional support to the viewpoint—actually a lie—of the first group of people. Sadly, what we can see today is that ignorance is overshadowing the ingenuity condensed in the three most authoritative law documents on this land. It is more and more incomprehensible about Americans' pride that they express when they say "No one is above the law in the United States." At least, in recognizing the absolute monopolizing status of English as an official language, Americans have allowed ignorance to take an upper hand above law, indeed, the three most authoritative bodies of law.

By demanding law stipulation to establish the official language status for English, the second group of people has been doing a wrong job. The monopolizing status of English as an official language in this nation has long been established by the Founding Fathers. It is up to the contemporary Americans to defend it or let it crumble. So the job of the second group of people is not to establish redundancy but to defend what has been there and theirs. Just simply roar: "Don't take our English away; he who intends to do so must be defeated and will have a high price to pay." Sometimes, because of the overflooding mentality of excessive democracy among some Americans, it is so sad to see the second group of people showing a mood nearly like begging in convincing people to use English. Why convince is needed? Why begging is needed? The Founding Fathers have given people enough legal tools to defend the throne of English in this country. Discarding English is an unlawful behavior on this land. A heroic father hates to learn to have a son who is lame; it is so heartbroken.

With the kindness of the parents, a child is not barred from learning other languages. However, when the family's affair needs attention, the parents permanently have the full right to demand the child: "Talk to me in the language I understand, in the language that you are raised." The parents never have to post on wall what kind of language it is; it is absolutely redundant. In this nation, nobody can change the fact that English is the sole language by which the Founding Fathers handed down the U.S. Declaration of Independence, the Articles of Confederation, and the U.S. Constitution. The Founding Fathers permanently have the right to demand the citizens of all generations on this land, their posterity, to use English—and English only—to communicate with them. It is true that their bodies of flesh are no longer with us, but what is more true is that we all are so fortunately blessed by their perpetual contribution to this nation that we must respect them as if they have never been separated from us. Communicating with them with a language that they declined to give value in a national affair is more shameful than disrespectful. Any attempt leading to the parting of English or lowering her supremacy on this land is beyond the matter of being constitutional or not, but literally a matter of loyalty or betrayal.

English is the silver spoon that this nation had in mouth when she was born. This is the sole language with which the citizens are expected to understand the nation's will, to obtain the nation's love and protection, and to understand each other's mutual pledge. This is also the sole language with which the nation understands her citizens. If a son is not lame, he will not feel the need to trade a treasure of inheritance of so many generations for anything. Whether or not American people can uphold English as their only official language is a critical test on whether or not America has the will and endurance to stay as one nation under God. If American people fail on this issue, disintegration of this nation will ultimately become inevitable. Americans, let no one yank the silver spoon away!

Genuine freedom leaves no room for freedom of betrayal. Now, citizens of the United States should have been warned of enough that both the Founding Fathers' faith and language are seen being trampled by attempts hostile to this nation. What is next in line to be attacked in the list of their tolerance? If American people cannot even defend their Founding Fathers' ideals, what do they mean to defend democracy and liberty?

Chapter 16

A Sinking Ship

In our daily life, we always hear of the term "American value." What is American value? Apparently, any human value that can be filtered as America's must be conformed to the nation's birth certificate: the United States Declaration of Independence. So any value that excludes the obligation of mutual pledge but declares as American value is a deceitful value—is a sham that is left over after castration. If this country continues to allow people to limitlessly demand "inalienable right" but requests no fulfillment of obligation from her citizens through mutual pledge, sooner of later, this nation will only be left with salvage value but not American value.

In 1978, flying the same socialist flag on the mast, China began to steer her economical vessel in a new direction. At this time, China was seen as dirt poor in comparison to Uncle Sam's treasure mount. In 2008, the year for a new American president to be picked, everyone in the world knew that China had become one of Uncle Sam's biggest creditors. There has been no war involved; what is seen on the political platform is only that these two countries joined each other to show a biggest economic joke in human history. No reasonable businessman will lose money to his competitor with such inconceivable speed like Uncle Sam. Isn't America the country that has the most Nobel laureates on economy in the world? If in the past thirty years this country can lose money at such an astonishing speed, what further assets will be lost in another thirty years? Can America afford it? The forecast is extremely scary!

Until 1991, the world in the twentieth century was dominantly influenced by three powers: the United States of America, Soviet Union of Russia, and Communist China. Western Europe should have also been one major power with her advanced economical ability. However, given that all the political and economical resources on this land are so scattering, so impossible to conglomerate into a concentrated force, Western Europe's influence in the world is quite limited.

Each of the three dominating powers influenced the world with a unique but outstandingly different "bankroll": Soviet Union with the most formidable land and thus the most abundant natural resources, Communist China with her tight grip on the biggest population in the world, and the United States of America with her bag that was so stuffed with cash that it would seem bursting almost anytime. However, time changes rapidly. While Soviet Union is no longer there at the end of the century and while China still retains her tight grip on one-fifth of the world's population, Uncle Sam is now seen asking China to lend money with a smile a little more dignified than a beggar's. With the "bankroll" so depleted, what chips does Uncle Sam still have in exerting influence to the world besides few carrier fleets that his grandfather left behind?

Democracy did facilitate the economic advancement of America in history. However, the advancement did not place America as the richest country in the world until about World War II. If we review the time line of the two world wars, we will find that American economy received a sudden forceful impetus during the period covering the two wars. The reason is clear: wars not only eliminated competitors in the world market but also brought Uncle Sam customers who found him as the only capable manufacturer.

Traditionally, Uncle Sam is not seen as a skillful lad in the modern international political casino. What supports his high-decibel voice there is his unmatched wealth. Very unfortunate to him, his unique political characteristic has been evaporating his wealth at an unmatched speed too. A typical figure showing the disappearance of his wealth can be reflected by the steel production. In 1900, America was producing thirty-seven percent of the world's steel while Uncle Sam was not even the richest country in the world yet. In 2007, China's production on steel is 490 million tons while Uncle Sam's trails behind with a figure of 97 million tons. In 1974, American steel industry employed 521,000 workers; in 2000, 151,000. How many breads and butter have become unreachable from the American dinner tables because of the change of these figures? At the time this book is written, the American car manufacturing, the traditional industry that has upheld so much American pride, is underway of reorganization through bankruptcy. Few years earlier, some economists did expect the looming of bankruptcy of the American car manufacturing industry. They put the year around 2020, but Uncle Sam "advanced" the time by ten years. If a grandfather had any feeling while lying in his grave, he must now be heartbrokenly crying in knowing how his grandson had been depleting the family treasure, losing his ability in the fiercely competitive world. To make it even sadder, while the financial crisis is cleaning up Uncle Sam's wallet, some Americans are seen extending their necks and looking across Pacific Ocean, trying the best to screen what difficulty China is experiencing. The mood is like "Ah-ha, you don't do any better either if I am in trouble." Frankly, China does have difficulty caused by Uncle Sam's money crisis; her difficulty is not having enough cash to complete her backyard patio. As to Uncle Sam's difficulty, it is that a big part of his roof is blown away while stormy rain is showering in. A long period of wealthy life has nurtured an unhealthy mentality: instead of figuring out the true reason of getting tumbled, one tries to look for relief from seeing if others would have to suffer the same.

Given how Uncle Sam has made himself rich, now Uncle Sam must rest his dream of regaining his wealth on that the world will repeat itself once more. This is to say that somewhere in the world somebody is preparing wars in big scale but not hurling Uncle Sam into it and that all manufacturing lines that once Uncle Sam monopolized come back to this land. What a romantic American dream!

It may be a little unfair to say that American politicians are completely unaware of the looming of some crisis. Before the apparent approach of the financial debacle, somebody has felt the pressure. However, given the resources that are truly under their control, the politicians have not too many cards to play with. All cards they play must obey one principle: make limitless promise of right to the masses. One of the cards is to put up the so-called stimulus package, one after one. Isn't a similar card found working before, such as Franklin Roosevelt's New Deal? Please, in Franklin Roosevelt's time, America had the most production lines among all the countries, and the lines were also the most rejuvenating ones in the world. During this time, America had more than abundant domestic cheap labor, while no compatible competitor was found in the world market; in fact, everyone else was too busy preparing war. Now, all these favorable conditions are either long gone or substantially thinned. Stimulus money can do nothing to restore the production line but only to reinforce currency circulation. Therefore, more stimulus money to a society that is devoid of production capability is just like more alcohol to a body deficit of bone marrow; the more vigorous blood circulation only speeds up this body toward death. Before the government distributes the stimulus checks, did any one in the government office ever speculate how people will spend the money? Be they a check of $600 or $250 for each person, it is not a big sum of money, helping nothing in a family but only satisfying shopping desire in the retail stores. If some investigation is done, such a government officer must have found out whose products from what country have occupied most of the retail shelves in the American market with unbeatable price. Let's look at some figure again.

From 1978 to 1988, China's international trading account balance fluctuated around the figure of 10 billion USD a year. An upturning began in 2001, the year a first tax refund check reached the American families in the twenty-first century. As the Americans' purchase power is gaining her might in the subsequent years because of the permanent tax cut across the board, China's balance account jumps from 17.4 billion USD in 2001 to 371.8 billion USD in 2007. It is a skyrocket up shooting of more than twenty times! Don't American people wish this to have happened in America? Truly, China did not make money only from Uncle Sam but from other countries too. However, the first few years of the twenty-first century are a period in which ideas of stimulus package flood the entire West or, more precisely, white majority lands including Australia and New Zealand.

Isn't the synchronization between China's skyrocketing figure of balance and the time it happens enough to tell people whose economy has really been stimulated with immeasurable impetus? If the Western politicians cannot detect the correlation between their stimulus packages and China's economic advancement, jobs found in other fields other than the political casino may be more suitable for their talents.

While damage done by the stimulus package is apparent in economy, the worse is yet to come. It is the damage developed on working moral. When the stimulus checks are pushed by political force to reach those who did not show willingness to work, the checks also send a strong message to everyone. Message to those who don't work is: laziness is not held accountable but awarded, and there is no need to pledge your fortune as required in the Declaration of Independence. Message to those who have worked hard or risked their investment is: if you had not been penalized, actually enslaved, by those who have chosen to be idle in contribution, your check should have had a bigger number when you receive it. Nevertheless, by giving everyone a free check and beautifying it as a stimulus package, a card that well matches the surviving technique in democratic political arena is played: to win over the opponents in election campaign is to win the rivalry of right promising.

So far, as this book is written, a report says that for every dollar that the American government spends, 47 cents in it is for interest. A loan with such an interest rate is throat slitting. America has not come out as a loser in any war in which she must sign to accept caustic treaties. However, she now has been seen acting in "full capacity" as if a loser out of a destructive war, reined by all limb-amputating terms: conceding land, paying huge wealth compensation to a victor, and appeasing. Land conceding is witnessed by tens of millions of illegal patriots flooding on her land. Wealth compensation is witnessed by the huge debt China holds against her. The report reveals that as of January 2009, China holds a debt of $740 billion against America. It is a debt of more than $2,400 per person. An American child, as soon as he/she arrives in this world, is made pledge a big fortune but to a country that his/her foreparents experienced chilling hostility. No baby can escape this debt unless he/she chooses to grow into a welfare recipient. ("Um, a welfare check from China must be bigger than the one from Uncle Sam," thought someone who is figuring hard what last name her fourth upcoming fatherless child should have.) What more horrendous is that the bill is seen keeping inflating, with no end can be envisaged, because what is seen is that the U.S. leaders continue to convince China for more "generous" loans. As to appeasing, does anyone still hear of any raising voice of human right from Uncle Sam toward China lately? Instead, the topic of discussion is concentrating on "green power." What a buddy!

Miracle is needed to restore America's halo in the past; many legendary heroes are needed for America to continue as one strong piece. Hope has not faded away for Americans, but time is certainly limited for them to survive as citizens with deserving pride in the world. Thanks to their foreparents' excelling intelligence and endurance, Americans still hold the most abundant and valuable real estates in the world. This can serve as their strong capital and "bankroll" for rebounding. However, they must realize that continuing to mortgage them is only to speed up their depletion. Backed by the strong real estate, they must think of ways to increase productions and quick. Without production, no stimulus money will really work, but all stimulus money they spend will only end up in other people's pocket. Without the sun, more vapors can only create more gloomy clouds but not rainbow. Yes, stimulus money can bring illusion of job creation for a short while. However, jobs created for one thousand tattoo artists or jobs created for one thousand workers next to a garment production line will make an immense difference to the nation's survival. Only jobs created in the production line make genuine money, and any job created elsewhere only dissipates money; this is plain mathematics.

America is a capital country. For the capital mechanism to properly operate, capital accumulation is a must. To make capital accumulation possible, American production lines in all industries must be restored, including summoning back those which have slipped away to overseas. To make the restoration possible, America must do at least two things without compromising. The number one thing is to break loose all social mechanisms that sternly aim at preying on the success of capital accumulation. The number two thing is to stop all social mechanisms that only serve to hide away the domestic cheap labor. Yes, given the situation that America has been in, these two things are extremely difficult to do, but doing nothing can only allow the quicksand to swallow up the nation at acceleration. It does require extraordinary genius politicians to lead America to race against time. The mentality of right promising in politician choosing must abruptly stop. The mentality of hatred against big business must be corrected. Here is a message for those holding negative view against big business: Aren't you glad that these eyesores have been leaving America one by one? Haven't their rapid disappearance from this land rendered you a cleaner landscape in your vision (so cleaner is your wallet too)? Don't blame their leaving; staying home only means to be preyed. Who wants to be preyed? Wasn't your negative view about them resulted from a feeling that you have been preyed by them?

A society of genuine democracy with true harmony will not tolerate either capital monopoly or labor force monopoly. A despotic society is built on the desire of monopoly of everything.

Thirty years ago, at the time China was about to decide to have her market open to the outside world, American businesses felt anxiously excited: With the formidable population found in China, even one tablet of aspirin for each Chinese meant a big business. Nobody ever thought of that China was then dirt poor; to buy an American aspirin, they must first make money. Nobody even had a slight idea that to make such money, Chinese workers were willing to ask for as low as 50 cents of RMB (Chinese currency unit, currently worth about one-seventh of USD) to make a pant that was about to sell for 10 USD on an American retail shelf. Of course, then, no consideration of what impact this would mean to the U.S. manufacturing was included in the calculation. China is not the only one country that has caused the evaporation of the American manufacturing lines but many others too. However, all these countries, including China, are not the ones to have opened the lid to let the evaporation escape but the Americans themselves.

If the production lines continue to evaporate from America, as well as from Europe, because of the preying by the monopoly of labor force, what Marx said must become true: "Its [bourgeoisie, naturally also the capitalist society] fall and the victory of the proletariat are equally inevitable." His "dictatorship of proletariat" will eventually set foot at every doorstep of all citizens. Marx, Lenin, Stalin, and Mao Tse-Tung—none of them ever lifted a screwdriver in their lifetime. But this had not prevented them all from labeling their genuine dictatorship with "proletariats"; not a single trace of embarrassment or shame is felt by them in so doing. After arrested by these "leaders" or someone possessing similar political carnivore nature, people who once felt unable to tolerate monopoly of capital by few must now accept monopoly of everything by even fewer. However, this kind of people can be found in the past and in the future; they are so willingly to repeat the same history whenever they feel proper for some short-term interest. Given the social condition of the United States that is unseen in other countries, such as the diversity of races, of beliefs, and of cultural backgrounds, flooding of guns among people, having become the richest country within the shortest duration at the youngest age of a nation compared to the others, when the "dictatorship of proletariat" shows up, there may be several of them showing up at the same time. Then, people, prepare yourself to be torn by several uncles besides the one called Sam, who usually appears during wartime, finger-pointing with mean-looking eyes: "I want you!"

In America, so many people have got used to such a mentality: demanding right, then more rights, as if the topmost obligation in one's life is to have a right and the topmost obligation of their government is to grant the right. More and more people seem to have lost the awareness of the social responsibility or even the responsibility to themselves. The habit of overspending is thus spewing out. "It is my inalienable right to borrow; you interfere with my human right by questioning my ability to repay. Are you discriminating me?" "It is my life, and why should you care if I go over my credit card limit?" Various financial speculations to satisfy these mentalities culminate without bound. Finally, with a financial debacle, the society tells people in America (as well as all the Westerns): your prefulfillment of obligation has never been adequate in supporting your limitless "right" in spending. So many people are caught unprepared. Well, bankruptcy, foreclosure on properties, cars being towed away, and unemployment all happen at once. Dreadful as it has been, however, the abusive right pursuers should be glad that it only happens in this way. If the same mentality continues but cannot be stopped and reversed, people must prepare some more devastating plague yet to come. Mother Nature's rigid mathematical rule of 1+1=2 and 1-1=0 allows no slightest dent. The more the human beings distorted this rule, the higher cost they must prepare to pay back. If no obligation is fulfilled in her bucket, don't even think of picking up any drop of "right" out of it. Otherwise, someone must fulfill a bigger obligation somewhere, some other time in the future. Mother Nature is lean and mean, huh? You bet she is.

Chapter 17

Obligation of Mutual Pledge

Surprisingly, a typical rival to America as China, her premier did give Americans a valuable advice in 2008: Americans have been abusively overspending. Possibly, giving friendly advice for Americans to change behavior is not his very intention but just explaining why he needed to be cautious in lending more money to America. However, on hearing this, some Americans did jump to the roof, "Why do we need the Chinese government to tell us what to do?" Possibly, the Chinese premier should have also included an ancient Chinese motto in the advice: Medicine for curing is always bitter, you American hardheaded lads! It is absolutely a tragedy that some Americans cannot understand that the time has changed; no one has made such change possible but they themselves, through their "nobility," innocence, and naiveness in the course of abusive right pursuing.

To curb the mentality of abusive right pursuing, to know that they have been seriously uprooting their national foundation by the abusive right pursuing, Americans need to do one thing: from now on, whenever they read the Declaration of Independence, the nation's birth certificate, they should read from bottom up, although each sentence still reads in the normal way. Doing this, Americans will not miss the extremely vital clause of utmost importance from this document: "And for the support of this Declaration, with a firm reliance on the protection of divine Providence, we mutually pledge to each other our Lives, our Fortunes and our sacred Honor." If they read from top down, with exaggerated satisfaction, many of them may feel enough to stop at the period right after "pursuit of Happiness." By not reading to the very bottom line, these people have actually castrated the fatal part from this document, converting this document into a carcass without soul in their mind. Then, so intoxicated, these people make themselves ready to join the children troop following the Pied Piper of Hamelin. Those pied piper leaders, those hardcore abusive right pursuers, are usually politicians of special agenda pursuers with special ambitions. If we listen to them more carefully, they chant the same slogans as those classic socialists did: "Give me power. I will lead you to get your rights!"

In America, no right should be and can be lawfully granted to anyone without this person's pledging if the Declaration of Independence is faithfully obeyed. No law, including the American Constitution, should be able to allow any American citizen being so contradicting to the Declaration of Independence that he/she shall have a chance to slip away from mutually pledging to his/her country fellows with equal weight. More often than not, it should not be that declining or omitting a certain right is to be unconstitutional to a citizen; it should always be that one is found slighting the mutual pledge required by the nation's birth certificate if he/she insists to have his/her right to overshadow the corresponding national obligation. Therefore, while the birth certificate can never be modified, the Constitution must expand to have terms to deal with such slighting.

A faithful obedience and commitment to the Declaration of Independence must require that anyone's collateral in a mutual pledge is indisputably made as heavy as the others' in weight; anything less is cheating. And the Divine Providence must see to it that the cheating be impartially corrected. Truly, this is easier to say than do. However, this should not constitute an excuse in the citizens' daily political life that some very fundamental steps are not to be taken, such as the following:

1. No political right that is supposed to be a privilege of a genuine citizen, including casting a ballot to vote a political leader or proposition, is automatically passed to anyone only because he/she has reached certain age. To earn this right, he/she must take oath in court or any equivalent office. Many countries have similar ceremony for the youth to be recognized as an adult, so why not America? The oath must include "We mutually pledge to each other our Lives, our Fortunes and our sacred Honor." Right is not to be given but to be earned as a consequence of fulfillment of obligation. This is how it is shown in nature; it should be the same in the human society. To earn the same rights as provided by the Declaration, the Founding Fathers must pay blood. What is so wrong to require a youth who will enjoy what the Founding Fathers fought for to spend a few ounces of energy to get the same thing? The oath cannot even be taken in such a loose manner: Some individual cites it, and the rest just babble, "Yes, I do" to conclude. Everyone taking the oath must handwrite this oath word by word following a leading voice; then cite the sentence loud and clear; conclude it with "Yes, I do. Help me, God!" and sign and keep the handwritten oath for life as his/her first political document without signature of a guardian.

2. Any immigrant who aims at the American citizenship through naturalization must do the same thing. Everyone who looks for the

protection and love of this country must be made to understand that this country has full right to demand every one of her citizens to devote to her this measure of love: "Love me or love someone else—not both, not either, not anything in between." Anyone's pledge must be held accountable; this country must monopolize the loyalty from a citizen who declares his/her love to this country and demands the love from this country at the same time. In so doing, the new citizen must be made specifically declare abandoning citizenship from any other country under oath. Words from the Declaration of Independence of America must be made so rigid that even diamond has no comparison.

Do we hear a stronger and stronger voice on reformation of immigration law in these one or two decades? Good news! This will give our lawmakers a chance to consider how to shut the door against the Fifth Columns from other countries to flood in through naturalization. In such consideration, any naturalized citizen found guilty of spying for their original "motherland" but against this country must be prosecuted with at least perjury and treason besides espionage and sabotage, punishable up to the maximum severity of law. All related trial will be held in military tribunal, not civil court. It is only fair because the suspect, under the camouflage of a civilian cloak, has caused suspicion that his action facilitates the military might for a political entity that presents threat to this country. If no such consideration is included, any so-called reformation, if realized, only serves as a helping hand to the enemies' dagger that already pointed at Uncle Sam's midchest in addition to all nooses that he has put over his own neck. Oops, mistaken! Shut the door? What are you talking about? This is definitely an American dream of excessive luxury. What more hid in the mind of a big number of the law-makers about immigration reformation is how to tailor the door to the maximal size. Be the comers Sixth Column, Seventh Column, Eighth Column . . . , they are never enough. "Come on in, guys!" shout those law-makers. "You need this land; we need your votes. Together let's make a perfect demonstration on the rule of supply and demand." To them, nothing is more important than the opportunity to maneuver the nation to become not Americans' America, but just anybody's America at will.

3. Principle of "hold accountable" must be enforced without dispute. Law—any law—is only as good as it is applied. Just because this country has the most complete legal system in the world does not mean the law is playing the role it should be in this country. To be able to escape from accountability is exactly what is behind the mentality of the abusive right pursuers. As their population is increasing, the old American saying

"Nobody is above the law" has become more and more meaningless. A supposedly law-abiding country is thus seen walking on a more and more unruly direction; a richest country, which supposedly has low crime rate, thus becomes a country where incarceration spills over rim.

4. The concept of "hold accountable" should be applied not only in civil activities, but also in all economic activities. Therefore, all officers above a certain level in a private business that openly recruits capital from the society, such as selling stock to get money for a company, must post bond with his family's property as collateral. The American government must see to make the concept "mutually pledge to each other . . . our Fortunes" make sense. Someone who is in charge in a position, private or public, that can have access to or can handle large amount of funds should also do the same thing. Something like Enron or the savings and loan disaster in the 1980s are not allowed to happen again. Any award payment to such an officer, such as salary, bonus, and dividend cannot be immediately one hundred percent cashable; but a portion of them must be put under probation for a certain period by some neutral institution licensed by the government. The money under probation can be released only after a certain period that his management is proven not having left behind damaging effect to the company. If the company shows sign of illness, the public can have a chance to hold his belonging and payment to compensate the loss he has inflicted. He who bankrupts a company that has ever recruited funds from the public, even one penny, must be made bankrupted. In this world, nothing is one hundred percent safe, but this way of holding a higher officer accountable for a company's healthy operation should prevent some wicked one from evil doing to some extent and give the public better confidence. The hatred from the poor to the rich should then have a chance of being lowered by some degree too.

5. To set up a hierarchy system of citizenship, such as first-class citizen, second-class citizen, so on and so forth—this is not a joke; this should be an effective way to safeguard the democratic society from being periled by abusive right whiners. We discussed how effective this system helped the Chinese Communist Party maintain her competence in ruling in chapter 11, part 2. By being properly modified, this system should help the democratic society stay away from deterioration because of excessive right pursuers. It is very true that "all men are created equal," but it should be even more true that it is up to each person to keep up with such equality. Nothing can be more apparent than the equality between two people must have been destroyed after one of them is raped by the other. Who said that these two people were not made politically equal

in America at the moment each of them was born? But one of them pitilessly destroys it in a crime.

So, instead of being pressured by the necessity of sending every trouble maker to jail, it is so much better to have a system in which people try to stay away from becoming a trouble maker as much as possible. In realizing such goal, the following system should play a good role: first-class citizenship is granted to a person who possesses education of certain level and above, such as high school graduation, and is willing to take oath of mutual pledge; second-class citizenship is assigned to a person who is free from criminal record but declines to qualify himself/herself to take oath of mutual pledge after entering adulthood; third-class citizenship is assigned to those who committed misdemeanor; fourth-class citizenship is assigned to those who are just released from jail. Correspondingly, only the first-class citizen can have the right to vote; a second-class citizen cannot vote although his/her personal activity will not be restricted; a third-class citizen must be restricted somewhat, and the fourth class must be restricted even more. All personnel locked up in jail by law are not considered citizens at all but just prisoners. All subclass citizens will be given opportunity to upgrade the citizenship class but only as approved by court and according to the person's behavior after a certain period of probation.

No citizenship class is linked to property ownership, race, religion, geographical origin, and lawful occupation but personal conduct. Each person's citizenship class must be made available in public record; no privacy reason can have such record hidden. That is what mutual pledge is for; mutual pledge will lose its significance if it is kept private from each other or among citizens. Can we imagine a couple keeping the oath secret to each other in a marriage ceremony? The society must give a citizen with higher class more favorable consideration over the lower ones, such as providing job opportunities in the public sector or private sector, getting into university, getting social financial help, or even getting a loan. Believe or not, such a system of citizenship class has actually had a long history of practice in the democratic society. For example, those who file bankruptcy are subclass citizens in the economic field; those who are put under home surveillance and required to report to a certain law-enforcing office at certain time intervals are subclass citizens in the political field. What is wrong if such a system is reinforced with more detailed regulation and carried out with more transparency? A more systematically and more vigorously classified citizenship hierarchy will make each social member know more clearly where he/she is at and what opportunity he/she can retain with a higher-class citizenship. If it is said that, in a tyranny society, only a handful of citizens of higher-class citizenship practice the dominance over a much bigger population of subclass citizens, the above system will provide the dominance of an overwhelmingly bigger masses of citizens over a handful of trouble makers. This system should have the function to maintain the highest genuineness for the concept that "all men are created equal" and inspire more citizens to live up to the equality. A benevolent society that swallows trouble makers' evil acts and renders them unrealistic sympathy is only gobbling suicidal pills. The sad part is that when the suicide becomes real, she will not receive any sympathy in return but swear or curse or condemnation.

If there is a political requirement for a citizen to be a political leader, there should also be a political requirement for a person to be a quality citizen.

6. A prisoner must be made to know that prison is not a good place to enjoy but a place to restore the sense of fulfilling obligation. American jails have long been psychotically called "well-guarded five-star hotels." In other countries, typically Communist countries, prisoners are a good source of labor of almost zero cost. Although democratic countries will try their best to avoid enslaving, it is certainly more mistaken if they exempt the trouble makers from arduous involuntary labor. Strenuous labor is the only way to tell a person that a jail cell is not a comfortable place to dwell and that the only way to stay away is to behave as a good citizen while one has the freedom. American jails always give some people illusion: you can get what you don't get in the outside world, such as medical protection; meals that are considered feasts in the third world; entertainment that you may not be able to afford at home; or, even "better," some wicked strategy to deal with the outside world when one regains freedom. The American jailing system has become an extremely heavy burden in social management by providing the prisoners with comfort. This model of management on correction is a typical sign of governmental incompetence, a joke in the world, a serious waste on human resource, and a catalytic chemical seducing social suicide in the long run.

To make all these weird things possible in American prisons is apparently because of many decades of negligence of duty by some prominent officers who are entrusted to take oath in protecting the U.S. Constitution. We will have more discussion later on this topic in paragraphs related to the weakness of the U.S. Constitution. Another reason that American prisons cannot scare away trouble makers is because of organizations of labor monopoly. Damage to the interest of monopoly of labor force is obvious if the prisoners are placed in the market to provide cheap labor. In order to stop the jailed labor force from entering the market, no excuse can show a bigger halo than human right. If the grip of labor monopoly cannot let loose, America's ever-weakening ability to correct trouble makers will only aggregate more and more weight to sink the American vessel. Instead of supposedly curing what ails the society, the American jails' system actually further sickens the society.

On the other hand, to use a prisoner for the satisfaction of personal purpose, either by the correction personnel or by another prisoner, must be made into a serious crime. A law and a management system must be devised to deal with it.

7. Schools must be made a place to prepare students for two things besides academic knowledge: (1) be a responsible citizen who must understand that national obligation perpetually has priority over personal rights and (2) Without the mutual pledge made by the Founding Fathers then, they won't be here to enjoy on this land now; so, without the mutual pledge made by them now, there will not be the same United States of America to exist in the future. Under the flooding mentality of excessive democracy advocated by abusive right pursuers, American schools are found preparing the students more often with the following ideas:

a. This country had a shameful history. So hatred against this country is well justified, and seeking revenge against this country is more of an honorable thing to do.

b. One's biggest priority and obligation is to seek for personal right and more rights.

c. No authority can be so supreme as to be able to jeopardize one's right seeking. Otherwise, this authority should be removed.

d. While right is imperatively emphasized, national obligation, such as protecting her with one's life, is not found being an issue. This has witnessed in many people's mind that, for example, privacy has far more value than national security.

e. As much as hatred is advocated, deed of justice and heroism done by this country is completely sidelined or even sneered. The bloody sacrifice that Americans made abroad, except during WWII, is always explained as the relentless interference in other countries' internal affair. Has any textbook and any school ever told the children that, for example, if America had not fought hard in Korea and Vietnam, even the soil in the front yard of America would have been reddened today?

f.   So no hero, besides Lincoln, is found to be mentioned with thanks in the war of removing the slavery system, although more than abundant heroes must have flooded in the history for the Civil War to be won.

g.   Instead of heroes to be discovered from, the entire white race is taught to be hated but only because of the wrongful doing by a handful of them. All races on this land are taught to be proud but white. White must learn to accept guilt.

h.   Instead of teaching the children to say thanks, compensation for suffering is taught to take first priority, regardless from whom payment is requested to be made, even from one's savior, helper, rescuer, liberator, or benefactor.

i.   Right has an uncompromised quality, and therefore, it must be extended to include everyone, even all trouble makers, even enemies whose aim is nothing else but to destroy this country.

j.   While the definition of "American value" is never found strictly defined, this term is always used as a magic veil under which values that had truly thrived America are constantly put in the list to be removed, one after another. A strikingly outstanding example is the heroic "defending" on religious freedom: "Remove the 'one Nation under God' from our children's pledge in school"; an unconstitutional act is interpreted as the only way to defend the Constitution.

k.   Compared to all these, it is only trivial to find that American students are all A students in pop quizzes, but not everyone are so bright in math. Surprise to see some of them so struggling with a problem like $(2.4 \times 3.06 = ?)$ at college entrance examination?

In short, the American schools have been formatted to prepare the younger generation with more demand on personal rights, more demand on excessive democracy, more hatred between races, and more hatred to the nation. More and more people are trained to feel that the nation has owed them too much, and subsequently, the nation will never repay them enough for what they deserve. As much as hatred against the government and between races is taught the extension of right or sympathy or excuse to cover the enemy whose only job is to destroy this country. This country needs no Soviet Union to advocate anti-America ideology; her school system has created her own propaganda teams who are doing a better job. When more and more of these youths with poisoned minds infiltrate into various leading offices, private or public, people must someday find that messiah Lenin has discovered a better site than Kremlin to receive the hailing but on this land.

American education must not only get back to the track to teach students to have love to this nation and to every citizen regardless of racial background, but must also have firm will against any culprit whose action only leads to sicken or even destroy this country. If it is important to tell a student what he/she is entitled to have, it is far more important to tell him/her who and what have enabled him/her to have it and how he/she can extend the greatness of this nation to the future generations. Every student should be found to have a soul that has an unshakable desire of entwining his/her fate with the nation's future other than whining for more ice cream and sometimes even comforting the enemies. A brilliantly intelligent head, Archimedes's head, just rolled to the ditch at the very moment the enemy gained access to his country. "History will not give us a second chance," said one of the top American government officers during the 1980s.

8. Racism, either the old form or the new form, must be stopped with equal strength by law in both directions, from whites to nonwhites or from nonwhites to whites. The term "racism" or "racist" must be strictly defined by law and indicted only by court and cannot be used in daily expression to frame a person. Accusing a person of being a racist without court sentence has the same nature as accusing a person of being a robber, a rapist, or a murderer without court procedures. A person is indeed a racist, regardless from what race, if he frames another innocent person as a racist without court procedures but is able to procure a certain benefits as a result of such framing.

Racism, as a legal tool to make someone liable to some damage, should meet the following criteria: (1) certain act can be realized by a certain race, but the other race is forbidden to do the same, and (2) profit is gained at the realization of such action by a member of the race that launches such action. It is up to the court to find out if racism has been carried out. A person calling another person racist without the court's verdict may make him a racist on the reason that criteria (1) and (2) are found to match him well. Any prospective criminal attempting to accuse an on-duty policeman of being a racist against him must produce evidence as strong as one with which a woman can accuse someone of raping her. Failing to produce the proper evidence, such criminal must receive additional punishment on top of the crime that he is charged. The free use of "racist" by anyone against someone else in this country has reached a scale that can be comparable to the political persecution once found in some Communist countries. In those days, if a person in those Communist countries was tagged as "counterrevolution," "antipeople," or "antiparty" by anyone who disliked him, such person was in hot water; if the accuser held direct power over this person, such person was politically dead. Court? Forget it! Should a democratic society tolerate such Communist culture?

Racism in the old form has been familiar with by everyone; it is typically a pressure from the whites to nonwhites. If the same political pressure is found but is applied in the reversed direction—i.e., from nonwhites to whites—is it or isn't it racism? But why is it not? Racism in a reversed direction is just as vicious as the old form but, if not extinguished, will bring stronger damaging momentum in ruining the nation. If this nation does not want to meet a fate of disintegration, stop the racism, both in the old form and the new form. All our ancestors came out from the primitive caves with rules of the jungle. This means that they were less educated with civilization than we are in the modern days. By practicing racism, they were doing the wrong thing without being fully aware that what they did was wrong. In comparison, if modern people know what is wrong but still insist to do the same thing, it must only appear more inexcusable, regardless the direction that the racism is practiced. Will stone throwing with the same sin wash away the sin of the thrower or make his sin ever heavier?

Old racism, being fully wrong, is motivated by animal instinct most of the time. New racism, however, can be additionally motivated by revenge. Therefore, its damaging momentum must be eventually many folds stronger than the old one when conditions mature. Racism of the old form culminates in the time that capitalism ascends while racism of the new form amplifies by strength at the time capitalism shows sign of being weakened. With the special rhythm in which the new racism and hatred echo each other, new racism will provide socialism with some particularly powerful political ammunition in this country. American people, beware! When this nation must face the fate of disintegration, anyone, no matter what race you are judged belonging to, can be a victim. Truly, racial prejudice is always encountered in our daily life. But racial prejudice must be distinguished from racism. Racial prejudice is only opinion, staying at the stage of belief in private sector. Racism expressed with public discrimination is to gain interest for a particular group of people. Prejudice is omnipresent and even appears between married couples, and between parents and children. What is to be watched out is to not let out the prejudice to the public and influence the masses.

Nothing can change the historical fact that the United States of America is established by Caucasians. Today, 230 years after this country is established, this country shows the broadest racial spectrum in the world. Everyone should be grateful that the Divine Providence has provided us with such a wonderful land to supposedly practice racial harmony; converting it to a land of racial hatred is against God's will. Many races, except the American Indians and the Africans, voluntarily came here as immigrants with a hope to get a life that they saw no hope to get in the mother countries. So, in those early days, they came here at the need as well as at the consent of the Caucasians who had settled earlier. With this condition in mind, civilization must determine that only one of the following attitudes is not ridiculous in accepting employment. With attitude one, a newly hired says, "Thank you for the opportunity of serving you, sir/madam," followed with a smile and a handshake that signify an enthusiastic service to be on the way. With attitude two, a newly hired says, "The reason I am here is because you need me. I will be unfortunately forced to make money for you. Say thanks to me." It can be true that some of those old-type employers may have taken heavy advantage of the newcomers disproportionably. However, an opportunity was an opportunity; this opportunity was obviously unfound back home at all cost. Indeed, while accusing the old-type employers of excessively exploiting, many of the accusers, regardless of the racial background, just repeated the same thing to the even newer comers when chance surfaced up. However, even with the idea of being heavily exploited in a new land, nowadays, people are still seen willing to pay almost any cost to get access to this country. In some cases, the cost is equivalent to a slavery commitment in the black market. When such opportunity is politically openly given, why an opportunity recipient must end up with an angry attitude toward the giver? While we give no excuse to the handful slave owners in the past, it may relax the slaves' feeling a little bit if we measure their hardship with the scale that those prisoners experienced in Gulag, Soviet Union.

Given that the whites pioneered the idea of liberating the slaves, given that they did have formidable number of people—a number far more than slave owners—to have shed blood in liberating the slaves, to view the entire white race as a criminal race being fond of enslaving is illogical. America is not yet truly civilized if no one in this country is willing to and dare to give credit to those who had their chest ripped open for the liberty of the others. Why a word of gratitude is so difficult to spell out? If the loss of time suffered by the slaves must be compensated by the government that liberated them, what is the price tag for the blood, limbs, lives, and disappeared families of the heroic liberators, and who will pay them? If we look at each other with more thankful feeling, we will also receive the benefit of living in a more harmonic environment. It is what Christianity has been teaching all the time. Isn't America blessed to have been established by people who held firm belief on Christianity? Hatred only gnaws people's heart, haunts people with despair, and parts brothers and sisters within a nation away from each other.

It has become a trend that more and more people of various backgrounds demand the government to apologize and even compensate for some historical events that happened on this land. Incredibly, the government that has no sense of competence does intend to kneel at each of this kind of demanding. It is truly beyond comprehension that people risk everything to come here from every corner of the world just for a life they found so impossibly intolerable that they feel the need to penalize this government. If they found themselves so elevated with justice, wouldn't it be nice that the same justice make them feel the need to thankfully compensate the other group of people, who have built this country with so much magnificence and been willing to let these justice people to share? Wouldn't it be even nicer that the heroism they displayed on this government was equally demonstrated back home so that the government back home was made kneel at them and these heroes had not had to come here to suffer? What bothers people's understanding is that the more these people feel the government on this land intolerable, the more their population increases, rapidly increases. Why has logic been so ridiculously inconceivable and inexplicable?

Just like a person, a nation must make mistake during her growth. However, in this world, only this nation has been so consciously and constantly putting up self correction; some correction may even appear extremely costly, such as the Civil War. Why must a benevolent government be seen so inexcusable; and the people receiving benefit from her cannot feel easy until she is punished? Why can't people think that we have all overcome insurmountable difficulties in history before we come to this point to own such a great nation? Why can't people think that the love we show to this nation can only inspire us to work together, instead of punishing the government, so that the same historical mistake will not repeat on this land? The way to stop the repetition is to defend it, but not to penalize it. This country is built on the base of "mutually pledge," penalizing the government is to penalize each other between citizens. Frankly, because of the "mutual pledge" principle, no government agent is empowered to expose this government to any punishment from any group of people in any manner for any historical event. Any government agent who thinks he could do so would have been acting with power that is not vested in him. Whatever he promises to the punishment can be and should be rescinded.

Carrying the case to extreme, if the government can no longer withstand but must finally crumble and disintegrate because of the "successful" penalization from each group of different backgrounds, will each of these groups really benefit from her disappearance? One suggestion to some groups of extremely low population: By the time this becomes real, no matter which group of larger population you choose to show alliance, you must make yourself an enemy of another group of compatible size in population. Your children and grandchildren can be easily forced to stand beyond the shield of the alliance on the one hand, and constantly stay in a magnet position to draw attention of bayonet from the hostile group on the other hand. Before you have to entrap your descendents in situation no direr than this that can be found, the best thing to do is to protect the current government when she still can protect you; this is the government from which you found so much profit that you cannot get anywhere else. Political myopia always costs dearly. If you are unable to stop those who cannot wait to butch the hen for eggs, at least do not add your strength there. If you want to trail behind them, longing to share a piece of the dead hen, with your population, you may very well only end up with being forced to sweep the feather while they are holding the drumstick. Blood and tear washing the Indonesian land but shed by ethnic group of minority have not been found dried over there yet in so many historical events aiming at expulsion. One of the big reasons is that, holding abundant wealth, this ethnic group of minority is not seen loyal to that country. Loyalty is not what you declare you have shown, it is what other people see what you have shown. Lessons, lessons of blood and tear, do not forget! Not in the sense to revenge, but in the sense not to foolishly repeat or even create a bed you must lie in. Do not fall into a scheme in which you serve as a pawn bringing the booty back to the big factions behind the scheme at the cost that your descendents will be thrown into hell. If you do not want your descendents' hands to mix with other human parts in a basket for exhibition, like those that were found in Sierra Leone, Africa few years ago, the long term benefit for you to choose is to side with people who genuinely know what gratitude is. Losing the heart to say thanks is step one for people to start a slavery path; being ill "rewarded," emancipators will not come any more.

This country has been truly too young, and is thus easily conceived as easy target by someone for some abusive purpose. In comparison, historical governmental mistake in some older countries will receive completely different treatments. In 1978, when China, a country with an age of several thousand years, decided to reform her economic operation, she needed to gather people's will for a new march. Then, without even an open apology, Chinese Communist Party asked people to forget all her bloody wrong doing in the past 30 years. Under the slogan "let us look forward," corpses that can easily be counted to 50 million disappeared from people's memory, just like a puff of a cigarette smoke gone into the wind, let alone any other inconceivable suffering. Indeed, many people, who were lucky enough to escape to other countries but got family members left behind and got killed in the past "revolution," immediately remitted money home to help rebuild "our motherland." Too many people held this view: "Mother did spank us, but a pious son should never blame her but only defend her with all compassion." Too bad, America, you are too young to be regarded as a mother of anyone. In fact, to a nice young lady like you, people seem never criminalize and demonize you enough because of the wealth you have been able to create at such a young age.

It is also true that sometimes Christians, whose population is so dominant in this country, work overboard so that love is even extended to cover culprits who aim at destroying this country. It should be realized that Christianity has been teaching on good personal conduct only and hardly on country administration. Christians do need to pay attention to how sympathy should be proportionate: when they pay sympathy to culprits, they should pay far more sympathy to the national reservoir that has been suffered from the damage made by the culprits. Do not confuse personal conduct with government behaviors. We shall have more discussion on this topic later. As to the American Indians, it is extremely unfortunate for them that when the whites began their colonization to this land, they had not yet formed a country. The necessary power with which they wanted to show their right of ownership of this land was unseen. While history cannot go backward, everyone, including the whites, must all know that they owe a thank-you to the American Indians, for so many people have believed that they found this land first. Had time permitted them to have formed an effective country by the time the whites came here, probably none of us is found being here except some diplomats. On the other hand, he who found a land first is not necessary to have ownership of the land; otherwise, the entire moon should belong to America now.

9. A new spending habit should be advocated in America. Overspending is a duplicate of the upside down mentality about the relationship between obligation and right in the financial sector. The overspending

idea floods not only in the personal level but is also seen as a habitual guideline in governmental operation. National budget with deficit is seen almost every year. If a department saves some money this year, this department is not awarded but penalized by having a shrunk budget next year. If the government saves some money, pressure will spring up somewhere to demand the government to immediately spend it "for the people." Capital accumulation is not seen honorable but sinful in both the private sector and public sector in a capital society; isn't it ridiculous? When financial crisis onsets, both the government and the civilians feel equally unprepared and shocked.

As America is getting used to this no-saving mentality and clinging to the overspending economic "principle," a top U.S. government officer is said to have visited China in the spring of 2009, demanding the Chinese government to spend more of what they have saved. The purpose of such demand is said to relieve the world's economy. It is incredible. The answer is "Spending and more spending are exactly what fails you, Uncle Sam! Why should we fail ourselves?" Of course, this officer would never have a chance to hear this answer; it is an answer they tell themselves between each other with voice-losing laughing in their office. Before this country can restore the effective manufacturing lines, more spending only allows other competitors to insert a hook deeper to the nation's throat. Everybody must have a habit to prepare something for rainy days; sunshine every day only means the ultimate coming of a devastating drought. Possibly, a complete tax-free earning on interest resulting from personal savings in banks up to a certain principal amount is a good start; so is a good start to award some government departments for their effort in not only saving money but also in achieving the same goal with less money. To borrow more money from China and to buy more products from them, with the hope that they will be able to lend more money, will prove to be an extraordinarily bright strategy that is full of outstanding intelligence for every citizen of China. Don't let up, Uncle Sam. What a human right warrior you have been!

10. Everybody knows that the current welfare system is a vast abyss for the national surplus, and that if this nation wants to stand up again with a rosy cheek, the current welfare system must have a revolutionary overhaul. If "Pursuit of Happiness" is replaced by only "happiness," or "potion of happiness" in the nation's birth certificate, America would not have been the America we know of today. The Founding Fathers must have been mistaken for this group of people who so desire.

The topmost damaging part of the current welfare system rests on the fact that it eliminates the source of cheap labor by buying them to stay home. The second damage is to expand the troop of fatherless children. Producing more children is one of the strategies for many welfare recipients to stay being qualified and to keep the welfare money coming. Children who are fatherless even make this qualification easier to attain; men are consequently released from their natural responsibility, indeed even being urged to be more irresponsible. When the fatherless children grow up, they do not complain about the irresponsible father but about the incompetent government for not having provided enough money to them. The damage on economy and morality to the nation is dreadfully immeasurable. The society needs to tell people, "The children whom you bring to the world are consequences of your enjoyment. Wipe your ass. If it is not you, nobody else should pay the bill for your fun." It is Mother Nature's rigid rule to have parents of any animal to take care of their own young; human beings as an entire living collection must suffer the consequence at high price for altering her will and rule. No human being is God; anyone pretending to act like God only creates disaster.

The third damage is that the current welfare system makes the unselfish contributors feel being mocked, cheated, insulted, penalized, and, worse, enslaved. All along, watching all the downhill rolling of the West, many people conclude that it is because the West has lost her will. It is hardly so. People of the West, at least of the United States, have hardly lost their will. You know they have strong will when you learn that the firefighters sternly and steadily ascended the stair in the World Trade Center on September 11, 2001, knowing a possibly irreversible disaster was just above their heads. You know they have strong will when you learn that the blood bank eventually had to set aside so much excessive blood unused after the Red Cross asked blood donation for those being wounded in the disaster. American people have never been highly disciplined by authority, but they often discipline themselves at their own will. To make all these possible is only because they have strong will. The welfare system must stop mocking and insulting those unselfish heroes—stop making them look like fools and being enslaved. The welfare system should still be there but only limit the help to cases like (1) temporary relief for those who get struck by misfortune that is really too big and beyond one's capability to salvage the situation; (2) assistance to offsprings of those, who lost their earning capability or even life in public service, such as soldiers, policemen, and firefighters, to enter adulthood or four-year college; and (3) help for someone who wants to improve his/her ability to achieve a better living through financial means such as grant, scholarship, or paid apprenticeship. Welfare money can even be used to guarantee a certain payment for a limited period of time that minimum wage cannot make up, if the recipient guarantees to work for minimum hours and to father no child soon. The overall idea is to let people know that the nation is not here to see anyone to fall but help him to get back to his feet. The nation will give an unlucky one a fishing pole, not a fish. The nation may even give him fish for him to get the energy for the first few days. However, if he must toss the fishing pole aside and lie down and whine for another fish to come, that is his choice—a sad choice.

11. The value of capital punishment must be confirmed. It is the only effective warranty in safeguarding the "pledge to each other our Lives." No concept of any kind should be allowed to stay above the national birth certificate so that the government's duty of monitoring equal pledge between citizens is jeopardized. If for any reason the government enables two individuals to pledge their lives with unequal weights, the government has acted not better than a cold-blood cooperator of the murderer.

That death penalty cannot stop murder is a "theory" that cannot withstand test of facts. China, a country that never hesitates to apply death penalty, executed 1,700 criminals in 2008. Let us assume that they are all murderers. Given that guns are not flooded in China, we can assume that the proportion between murderers and victims is one to one. So 3,400 people lost lives because of murdering. Contrary to this figure, every year America loses about ten thousand innocent lives because of crimes. This is to say that if America has a population like China, she will have more than forty thousand of her innocent citizens losing life yearly. Yet only thirty-seven criminals were executed in 2008 in America. If someone says America has made her land a paradise of murderers, what is the counterargument? If someone says that an innocent life has higher potential to disappear in America than in China, how would America demonstrate that she has had a better way to defend human rights? With the sharp contrast of the number of victims of murdering between China and America, how can "Death penalty cannot stop murder" be concluded? The way America applies capital penalty can be interpreted as that someone does encourage murdering. Worse yet, even when capital punishment is applied, the criminal must be watched to make sure he dies in the least painful way. Opponents against death penalty say, "Each individual has a right to life, even those who commit crimes such as murder, and if we take their life from them, we are no better than they are." Really? So is a government better if she acts like a cold-blood cooperator of the murderer to enable more innocent lives to evaporate? All we can say to these opponents is that any individual committing murder has surrendered his right to life at the very moment that he causes the loss of life of the other. That is what mutual pledge between citizens is about. The mutual pledge has enabled the democratic government to form; therefore, the mutual pledge must force the government not to deviate from her duty and see to it that the murderer necessarily loses his life without other choice. Since currently there is an argument saying that lethal injection is not necessarily a way to die without pain, then, let the murderer choose some way he thinks more humane to him: fire squat, electric chair, hanging, guillotine, or in the same way that he has made the innocent one suffer. If a murderer's life is so elevated in value compared to the life of the victim, if a murderer is so highly guarded compared to the victim when he/she still owned the chance to walk on the street, it is definitely surprising to learn that the heroic opponents against death penalty have not yet listed murder as one of the human rights.

12. The law enforcement agents, who help remove the internal trouble, and the military soldiers, who help remove the external trouble, must be buttressed with value, respect, morality, and competence. There is no single reason why any trouble maker, internally or externally, has higher value attached to his life than the life of these agents and soldiers. These agents and soldiers are the power to guarantee that the pledge required in the nation's birth certificate can be genuinely mediated. To an internal trouble maker, his wrongful behavior must have devalued his pledge; to an external trouble maker, not only his pledge is not found, but he is also found only intending to destroy the nation. Therefore, in any measure, the Declaration of Independence has empowered no one to place a trouble maker's life in a more valuable niche than the agents' and the soldiers' life when they faithfully carry on their duty. To place the life of the trouble makers in a more valuable position matches so well with the interest of the enemy but violates so much against the interest of the people who mutually pledge to each other.

Overall, the nation's birth certificate has not allowed any abusive right to be granted to anyone. It does not grant anyone any right with any nomenclature of enjoyment that can be possibly invented. On the contrary, it does require that any right to be granted must be fully countermeasured by fulfillment of corresponding obligation. Any so-called civil right or human right is actually an open infringement of the value of mutual pledge if it only ends up with having another fellow countryman to become a victim of such "right."

—

America is the greatest country in the world. However, facts tell us that she has been enticed to a place where certain reviving remedy is critically needed. The work for reviving is formidable, and the action cannot be waited, in all fronts. If in only thirty years, this country has lost so much wealth to another country, can she afford to let go another thirty years?

Just how long is thirty years? In 1991, the Soviet Union disappeared from the world. However, only thirty years back, in 1962, her power gave her so much illusion that she felt strong enough to intimidate the United States at the Cuban crisis. In 1914, the year WWI started, the United Kingdom held the biggest empire in the world; wherever the sun moved to, it must shine on a British flag. A little more than thirty years later, in 1947, she prepared herself ready to accept Marshall Plan in order to stand on her feet again. In 1917, the free world lost a country with the biggest land to Communism; it took only another thirty-two years for the free world to suffer another huge loss: A country with the biggest population was sacked by Communism. In 1945, the United States was an absolute victor in wars in both hemispheres; in 1975, exactly thirty years later, her soldiers came home from Vietnam with a blank report card, unable to declare as a victor. And following that, only another thirty-three years later, a dirt-cheap country became her biggest creditor. Even more miserable, counting from the year that America was said to be the only superpower in the world as Soviet Union dissolved, it took only seventeen years for her financial power to be linked to the mercy of another country. All these are to say that if a country does not take good care of herself, thirty years are enough for her to have a potential to disappear, regardless how strong she once is. The astronomical loss of national wealth is only an alarm to the Americans. They should still feel lucky to have such a forewarning. If nothing can be done properly on time, shroud making may be the only manufacturing line to be found left on this land. It is not always a good experience for some "first time" event: Through a binocular, the first time ever in this country, standing on a rock at a beach, an American kid detects a flag that he is not familiar with on the masts of a carrier fleet, which seems not hurrying to anywhere and taking its time.

Clearly, soul, language, and wealth of this nation have all been seriously threatened. Carrying the case to the extreme, if soul, language, wealth, and education are all found removed and successors are discontinued, what alternation does a country have besides disintegration or being conquered? As all these losses are happening in America, everybody in the world is predicting that America will lose the global influence to China. With signs that have been seen, if the ultimate loss is limited to only the loss of the global influence, it should be considered a good thing to have happened to American people, because, at least, the country still retains her completeness. Few paragraphs back, we mentioned that one of the top American government officers told people in the 1980s, "History will not give us a second chance." With the debacle of financial loss so overwhelming currently, America so far still stands firm; God does have given America a second chance. Folks, treasure it! To expect a third chance may just expect to be wiped out.

> If evil of this magnitude can be ignored, if our own children forget
> then we deserve oblivion and earn the world's scorn.

The above is an inscription found below a huge cross on Mount Davison, San Francisco, in memorial of the Armenia victims of genocide in 1915. It is certainly a sincere hope that no ethnic group of any background must repeat this heartbroken mourning to their children in the future only because this group was once American citizens. When some less misfortunate Armenians were able to escape the genocide, they found a land in America and laid down this inscription. In case some ethnic group must escape the scorching that sweeps America in the future, no land anywhere in the world will be available for this purpose. Waiting to see an arrogant and ignorant wealthy boy going broke satisfies many people's feeling. Besides, being a pioneer and warrior for democracy and freedom, America did have a history of offending many thugs in the world. And, unfortunately, all thugs seem to have more chance to get descendants and descendants who know to stand on their fathers' side too.

Chapter 18

Weakness of the U.S. Constitution

a. A submissive position

As the seemingly most authoritative document that law workers on the land of America must always and, so far, ultimately refer to, the U.S. Constitution has many weaknesses.

The number one weakness is that she must be submissive to the Declaration of Independence of America, regardless how superior she can be to any other documents on this land. Without the certificate, the U.S. Constitution should have zero chance to exist. A judge, whose power comes from the U.S. Constitution, can render a verdict regarding whether or not a certain act or speech is constitutional or unconstitutional. However, what if a verdict is found not matching up with the nation's birth certificate? With the indisputable utmost grandeur of the nation's birth certificate, if contradiction must be ironed out between this certificate and any law document on this land, the nation's birth certificate must stay unchallengeable.

A very typical example is seen in the case of the removal of a 2.6-ton granite monument in the Alabama state building in November 2003; on that granite monument are inscribed the Ten Commandments. The reason with which the judgment is made to have the granite removed is that the existence of such granite monument is found in violation of the U.S. Constitution's principle of separation of religion and government.

First of all, directed by such judgment, one must feel puzzled, baffled, and surprised if he/she ends up unable to detect the principle of separation after he/she reads the Constitution line by line, from top to bottom, including all the amendments. If no direct words can be quoted for such principle, then this "principle" is exactly nothing more than a politic creation of someone else and absolutely not the belief of the Founding Fathers. Contrary to what this someone "found," indeed, it is the U.S. Constitution that tells people that *our Lord* witness the date the Constitution comes into effect. How has this witnessing been a principle of separation of religion and government?

Secondly, once God's words of Christianity teaching are found attached to a public property owned by the nation at any government level, an action to have such words removed is equivalent to an action to have *Creator* and *Divine Providence* chiseled away from the Declaration of Independence of America. How could anyone find that he/she has been empowered to apply the Constitution's "principle" to the Declaration of Independence? Why a governmental decision with bias to non-Christianity beliefs over Christianity belief is more matching the "principle of separation of religion and government"? How this decision of obvious law making has not contradicted "shall make no law respecting an establishment of religion, or prohibiting the free exercise thereof"?

Thirdly, if someone is so constitutionally conscious, he must be fully aware of that he is seen as one of their posterities by the Founding Fathers, who regarded themselves all coming from one Creator, as so recorded by the Constitution and so stressed in the nation's birth certificate. Any one of the top officers who are entrusted to run this country is presumed and allowed only to guard what the Founding Father handed down to him. Yes, the Founding Fathers do allow freedom of religion and require no religious test as a qualification to an office. However, the togetherness of freedom of religion and the word "posterity" in the Constitution signed solely by Christians only assumes that he who has not openly declared as a non-Christian on this land is presumably regarded as a Christian. It is absurd that when a man passes his property to his son, he needs to verify his son's family name beforehand. If a person openly assumes the status of a Christian but acts against the belief of Christianity, how is he supposed to resist the suspicion that freedom of religion has been converted by him into a freedom of destroying that aims at the Christianity belief? When the Founding Fathers allow people from other belief to take care of this country, it does not mean that they allow such person to change course of this nation to somewhere they disagree. Can we imagine that any of the Founding Fathers would have agreed to have "In God We Trust" changed to "In Allah We Trust" or "In No God We Trust" in the back of all American monetary bills?

Indeed, given the historical background, when the Founding Fathers required no religious test, they may just have limited this no-test requirement in their minds to all the branches and denominations but within Christianity. As to any other religion, which was another term to mean faith, outside of Christianity, theism or atheism alike, they just regarded such religion being trivial on this land. Such limit is proven as a historical practice in their minds by the words found in article III of the Articles of Confederation:

> The said States hereby severally enter into a firm league of friendship
> with each other, for their common defense, the security of their
> liberties, and their mutual and general welfare, binding themselves

to assist each other, against all force offered to, or attacks made upon them, or any of them, on account of religion, sovereignty, trade, or any other pretense whatever.

In this article, religion is presented as singular, not plural. Of course, then, what the Founding Fathers would not allow on this land is the existence of any force or attack against the sole religion of the Founding Fathers', but nobody else's. In the above quotation, if the word "religion" is replaced with the word "Christianity," the spiritual principle of the document can remain intact. However, if the same word is replaced with "Islam" or "Buddhism," the document must tell people that all those Christian signers did not know what they were doing, and the entire document must all fall apart.

The above quotation appeared seven years before the Constitution was established. Eight years after the Constitution was established, George Washington told the nation: "Reason and experience both forbid us to expect that national morality can prevail in exclusion of religious principle." Words were thus spelled out by the nation's first president who assumed the office with one hundred percent of the votes of the Congress. Will somebody be bright enough to teach the father of this country the "principle of separation of religion and government"? Will somebody tell the nation that the religious principle in Washington's mind is anything else but exactly not the belief of Christianity? Therefore, in the process of establishing this country and, in particular, before and during and after the establishment of the U.S. Constitution, the Founding Fathers also established a principle that the U.S. government cannot separate herself from spiritual support of religion, which can be found only in Christianity. The quality of the inseparability is just like blood and live flesh. Only attempts serving betrayal purpose, intentionally or unintentionally, welcome the "principle" that government and religion must be separated on this land, or even the claim that this country must be a non-Christian country. Indeed, partisan politics, which is opposed by Washington, is by itself a practice in which religious principle, although atheism, and government mingle together neck to neck. Can anyone separate them?

With the three documents so constructed and so signed on this land, religions on this land are assumed by law to be categorized into only two: Christianity and non-Christianity. If the concept of "no religious test" must be extended to embrace beliefs outside of Christianity, law makers and law professionals must let either Christianity or non-Christianity prevail; there is no other choice for them. In the name of religious freedom, a legal action leading to officially suppressing a person's Christianity expression is openly escorting the non-Christianity belief with governmental power. It is a naked violation of the "no religion test" requirement if it is exercised over a person assuming office. Doing so, someone is actually destroying the Constitution in the name of carried-on fidelity of the Constitution. What is the definition of "perfidy"? Suppressing one's Christian expression is no different from the "don't ask, don't tell" practice regarding the homosexual activity in the military force. Not until now do people learn that Christians, therefore the Founding Fathers included, have been ranked with homosexual lovers to enjoy the same quality of pride in this country by some government agents. What a "favor" it has been! Suppressing one's Christian expression in this country with governmental power is allowing someone to declare with pride that "Nope, I am not a Christian"; but the pride of the posterity of the Founding Fathers is officially slaughtered. This means that even George Washington, if he is still with us by now, must keep quiet about his religious principle, or his honor may be striped off. Obviously, Constitution has given no one such power.

Another example reflecting the contradiction between the nation's birth certificate and the Constitution can be found on the issue of citizenship of multiple nations. With the Constitution so written, with even the term of "treason" being so loosely defined in it, an American citizen will find not much restriction from the Constitution if he/she desires to own multiple citizenships from many countries including the one from America at the same time. However, the mutual pledge requirement from the nation's birth certificate must forbid such desire from being realized. Anyone owning a citizenship from any other country besides American must have dented his pledge to the fellow countryman of this nation; there is no room left for excuse. Can a man or a woman belong to two or more families at the same time but pledge one hundred percent of his/her contribution to each family?

By allowing possession of citizenship of multiple countries, the government encourages a mentality of some species of crab spider-lings out of few of the native born: suck the mother until dry and leave. It also encourages a vulture's mentality out of some immigrants (of course not all, but still bad enough to the country): swarm, peck, and fly away. The "love" from both groups to this country will stop at the point that she can no longer stay as a meaty carcass. Given how young this country is compared to many other countries, given how diversified the human background in this country is found, allowing possession of citizenship of multiple nations is an extremely foolish generosity in political operation.

It can be said that America has become physically strong, but as a nation, she is still too young, miserably immature in personality and idea. She did begin her journey with a good start toward a good direction. However, she still needs good and consistent self-disciplinary action to guarantee her to walk on the same expected path without faltering. Allowing multinational citizenship only retards her from growing into a more mature personality as a nation. A physically full-grown and vibrant young lady is put to work unbelievably hard at very early adolescent age even before she fully learns about this world; but she is still also tempted and forced to share all what she can discover and create in order to make everyone, including the enemy, happy. What can be sadder to happen to this person during her growth? What can be a bigger sin and crime that are committed by those iniquitous factions surrounding her? Extortion from the United Nation toward the United States is a typical example how America's kindness has been ill exploited. "America, if you want to advocate a noble idea, then do something noble: Pay for this microphone so that I can rip you off in the UN!" If America wants military backup for the noble goal, too many of them will either stay away or even stay at the same front with those countries that resist America's effort. Unexceptionally, however, they must all drown America with the swearing of imperialism. They even put up this question: If America can have nuclear weapons, why can't the other countries? If a policeman has a gun, it is only fair that a criminal will have at least the same. Political nature and principle that each country stands for is not a concern; what they are concerned is "fairness," and through this "fairness" America can be reined by them. Didn't we hear some Communists in the UN meeting insist that the world has to be a world of multipoles? Isn't it nice that the same people also allow this kindness and enjoyment of multipoles for the common people in where they have ruled? Set aside how America's enemies have pitilessly taken advantage of her innocence, even some peaceful immigrants to this country only regard her almost like a province of the "mother country" that they came from. With the money they make in this "province" and ship out, they profusely earn praise of "patriotism" from back "home," but in the honor of their "motherland," of course. America has come to a critical point to review her generosity on the issue of citizenship of multiple nations. Mutual pledge in the nation's birth certificate are plain words to forbid the owning of citizenship of multiple nations by anyone who owns one from the United States of America.

b. Selectivity in Application of Law

The U.S. Constitution allows selective application of law. At where selectivity is severe, it even allows terms lacking strict definition but loosely gauged, such as human right or civil right, to prevail. A very typical example is the practice on Amendment 13, which reads:

> Section 1. Neither slavery nor involuntary servitude, except as a punishment for crime whereof the party shall have been duly convicted, shall exist within the United States, or any place subject to their jurisdiction.

> Section 2. Congress shall have the power to enforce this article by appropriate legislation.

Apparently, this amendment demands the necessity of punishment up to slavery "for crime whereof the party shall have been duly convicted." Amendment 13 is adopted on December 6, 1865, more than half a year after the final day of the American Civil War, which officially ended on April 9, 1865. Therefore, abolishment of slavery on civil land cannot be an excuse not to put criminals in arduous labor on land designated for correction if this amendment is fully respected by all branches of the government. Furthermore, section 2 even demands the Congress to "enforce this article by appropriate legislation." So it is not that putting criminals to arduous labor violates human right; it is that failure to enforce this amendment should have been found unconstitutional. As the government allows the American prison system to develop into a situation as found today, it is absolutely a consequence of long-time negligence or even contempt of someone's constitutional duty. The lawless illusion developed by the selective application of the Constitution even allows the extremely flexible concept, the so-called human right, to override both the Declaration of Independence and the Constitution in many situations.

Today, selective application of the Constitution is even on the way of creating and enforcing new racism. Such creativity and enforcement are typically seen in that nonwhites are allowed to form an organization based on their genetic characters, but whites must be forbidden to do something similar. On this matter, the U.S. Constitution fails herself big time if someone ever aims at the removal of racism by means of the Constitution. The new racism that the Constitution enables to indulge is fully soaked with hatred and revenge. Instead of securing the well-being of America, the Constitution with some of her weakness actually allows someone to incessantly plant timing bombs on this land. That "you are the beneficiary" must be read as a ghastly message by some people. To them, this message is conceived as that some folks will not stop at what they have freely received but are very focused at complete redistribution of possession. Two opposite concepts thus tensely stare at each other. The only legal tool to stop these two distantly remote and opposite concept from further parting is the "mutual pledge" requirement from the Declaration of Independence of America; the Constitution has not been seen serving to calm the ever-vigorously surging tide.

The selective application of law that the Constitution fails to stop even allows excessive democracy to be more powerful than military force. For example, Amendment 3 reads:

> No soldier shall, in time of peace be quartered in any house, without the consent of the owner, nor in time of war, but in a manner to be prescribed by law.

However, a property owner must allow a failing tenant to continuously quarter in his house without paying rent for as long as 6 months in some cases in some city and cannot recover his property unless he can afford additional loss to hire a lawyer. Will the tenant be held accountable if he happens to be so wicked and purposely inflicts damage to the property when he is finally forced to leave? Good dream!

c. Wording Vagueness

Wording vagueness is found. It is this reason that some amendments are felt necessary to have been introduced to clarify such vagueness. However, some other vagueness remains. A typical example is found with the First Amendment regarding the religion, which reads, "Congress shall make no law respecting an establishment of religion, or prohibiting the free exercise thereof." It has not specified if the reason to "make no law" is because of intentional declining for a purpose or because making law on this matter is unnecessary (for the reason that Christianity has been so overwhelmingly held by the signers). By writing in this way, competition between religions is fully encouraged. Therefore, the Constitution allows room for religious persecution to wait for the maturity of some political conditions. "Merry Christmas" is now seen semi officially suppressed in public areas for greeting but replaced with "Happy holiday" in a country where more than seventy percent of the populace holds the belief of Christianity. A nurse praying for a patient with words from Christianity is risking her job in the hospital. Seemingly fair, these practices are actually the result of silent but unfair encouragement from the Constitution because of her wording. Because "Congress shall make no law respecting an establishment of religion," non-Christian beliefs, theism or atheism, feel they have found governmental support in taking away the Christianity value. When a belief must suffer because another belief is inspired at the support of political power, a sign of persecution against this religion begins. Freedom of religion is obviously enjoyed more by a certain religion over this one that must suffer; freedom of religion must be felt as an open lie by this suffering one.

On the other hand, in fact, it is exactly because the First Amendment is so worded that the Christians have the full right to reclaim what they have conceded. They are backed by Creator, Divine Providence, and our Lord that are put in the most vital national documents by the Christian Founding Fathers. In the Articles of Confederation, the Founding Fathers expressed clearly that the sole religion they hold is also one of the reasons for them to assemble and form the country that is called the United States of America; the religion they hold tolerates no force or attack against it on this land. What is so wrong for the

---

prosperity of the Christian Founding Fathers to reclaim their home ownership of this land when Congress makes "no law prohibiting the free exercise thereof"? It is unbelievable to see Christians are so willingly to accept defeat when the most parental law bodies of the nation are so favorable to them. All it can be said is that excessive democracy has dwelled more and more firmly on this land, making more and more people feel losing their mind.

Christians seem not yet aware of how powerful they can be if they unite and just set aside the trivial difference between them. All the differences between them, such as how a certain ritual should be, are only ideas bound by humans' simple mind; what is common and essential to all of them is how to be a good person and devote wholeheartedly to God. All Christians should stand together to tell some authority: A nurse has a full right to pray for a patient with words from Christianity teaching if the patient expresses the willingness of acceptance. Any person stopping such nurse from so doing violates this nurse's constitutional right that is decoded from "Congress shall make no law . . . prohibiting the free exercise thereof." On this land, every Christian has full right to assume an unfamiliar person to be a Christian as such a person has been presumed posterity by the Christian Founding Fathers through the Constitution. Therefore, unless this unfamiliar person openly declares his/her religious belonging in one way or the other, a Christian clerk has full right to greet a customer with "Merry Christmas"; it is up to the customer to respond with a choice from "Merry Christmas" or "Happy Holiday" or silence or even "Allah is great" or "Nan-Mu-O-Mi-Tuo-Fuo" (a Buddhism greeting). It is not up to an authority to unconstitutionally threaten this clerk with his/her employment.

In the American daily political life, people encounter more and more frequently the words "conservative" and "liberal" in place of "Republican" and "Democratic" as the mainstream labels for identifying people's political stands. The conservatives certainly have missed the Founding Fathers' profound wisdom big time if they fail to support the Christians to reclaim their home ownership on this land. If they continue to be ignorant about the immense political might of the union of Christianity, the agendas of both the Conservative and Christians must inevitably meet their end of failure. No other choice is available to them. The Christians have been anxiously longing for the homebound journey of the 2.6-ton granite tossed away from the Alabama Federal building for many years. Are the conservative politicians anxious enough to lend a helping hand with the legitimacy that they could tab from the three grandfather documents? This is a critical test of how much determination American people still have remained in protecting the entirety of the Founding Fathers' ideals. If they are willing to fail this test, America, forget about your continuous existence. Forces that aim at dissolving this nation have been allowed to secure their potency in changing the course of this nation.

There are some other places where the matter is even beyond vagueness. For example, treason is so mentioned in article III: "Treason against the United States, shall consist only in levying War against them, or in adhering to their Enemies, giving them Aid and Comfort." As such, espionage against this country is found being the low-cost but most profitable business in the world during "peaceful" time, as this country has the most desirable secrets in the world to steal and sell: military, technology, and commercial. More than plenty of these spies are holding the American citizenship. Neither any word from the Constitution is found regarding how to stop a willful politician from selling citizenship together with terrestrial habitat and natural resources to foreigners at the expense of the nation's destruction and in exchange for personal prosperity.

d. Absence of Bonding between a Nation and Her Citizens

The Constitution has very detailed stipulation regarding the obligation and right of all the government branches. However, the entire text body of the Constitution, including all amendments, has no direct word regarding the obligation of a citizen to the nation. All obligations of a citizen are read in the Constitution as a result of right exercised by the government. There are two very negative consequences in such a way of wording. The first negative consequence is to take away the natural bonding between a citizen and the nation. The second negative consequence is to distance the government and the citizens away from each other as if they had been two political factions of absolute opposite interest. One can read the spirit of "government of the people" from the Constitution, but one cannot read the spirit of "nation of the people" from the same document. Therefore, love from the citizen to the nation is made "legally" absent, although the government established, according to the Constitution, is supposed to be "of the people, by the people, and for the people." Therefore, so many people in this country are found holding so much negative view about their own government, although this government is, in fact, the most benevolent government to her people in the world.

A contract that has no obligation but only right is a fake contract and, indeed, in many cases, is extortion from some powerful party over some powerless ones. With the citizens' obligation to the nation so feebly mentioned, but their rights so conspicuously emphasized, the U.S. Constitution puts itself in a position nearly like a fake social contract, leaving unlimited room for people with special ambitions to extort the nation. "The powers not delegated to the United States by the Constitution, nor prohibited by it to the States, are reserved to the States respectively, or to the people," so states the Tenth Amendment. So paying tax by a citizen is almost like a consequence of persecution from the government while abusing or even willfully damaging public facility is almost like a constitutional right. At the absence of clauses about military drafting of healthy man at certain age and mandatory wedded reproduction by healthy woman at certain age, these two natural obligations to a nation are again seen as unreasonable political pressure from the government. The consequence of such ignorance found in the Constitution is to have young men consider a political candidate ill qualified if such candidate proposes military draft and to leave pro-life and pro-choice fiercely fighting even with blood spilling.

e. Absence of National Principle

The American Constitution fails to stress what kind of social nature this country should stand for. If it is not because of the Declaration of Independence of America but just by herself alone, the American Constitution even has a very low voice to announce this country being Christendom. This is a dangerous pitfall of negligence that is excusable to the Founding Fathers, but inexcusable to the modern Americans.

All along, ever since the Founding Fathers, people just assume this country to be a capitalist country. As a capitalist country, the vital natural marketing rule of supply and demand is assumed to be there enjoying imperative supremacy and governing the capitalist operation in people's daily life. In the Founding Fathers' era, Socialism may be an unfamiliar term to everyone, and the marketing rule of supply and demand had not received thorough discussion like today. It was so natural for the Founding Fathers not to stipulate in document what kind of society this country should realize besides religion and language, but just to assume that transaction between buyers and sellers with full willingness in the market was no more than natural. However, modern Americans must be aware that people worldwide have been equipped with a completely new set of social knowledge. With this knowledge, Americans must fully realize that they have been besieged by an overwhelming danger of Socialism, external and internal, and that the marketing rule that supports and governs the capitalist operation has been ever increasingly distorted, deformed, choked, and amputated. With the marketing rule being clipped off, so will their democracy and liberty. If the marketing rule can be "lawfully" removed from the society, sorry, the same American Constitution can serve the socialists equally well. Fellow Americans, the kindest and most generous people in the world, please read your Constitution when time is still with you:

Article II—The Executive Branch
Section 3—State of the Union, Convening Congress

He (the president) may, on extraordinary Occasions, convene both Houses, or either of them, and in Case of Disagreement between them, with Respect to the Time of Adjournment, he may adjourn them to such Time as he shall think proper.

Congress has nothing to restrict the president's power when the "Time of Adjournment" is thought proper by him. All he needs is a longer term presidency, such as twenty five years, or "better", with tenure. People, are you surprised that illegal immigrants keep pouring in to this country without any barricade possible to be put at the border? Besides the economy's need by the combination of some business owners and factions of monopoly of labor, someone needs to alter the social structure at the bottom level. The illegal immigrants are so well protected, so well pampered with all kinds of benefit; they are not fed with free lunch, but they are needed. Their head counts are needed by someone who feels so tempting to have this country changing hand someday! A cunning long-term political calculation has been formulated with all the slyness; do people need to be convinced that a conspiracy has been carefully put underway? Any constitution in the hand of a monarch of absolute power is merely a piece of paper that is immune of any ink but his.

Changing terms of the president's office has happened to the Constitution in history, why such history cannot be repeated but just with a different numerical figure? Papers spewed out from the ballot box can do the job just that. American people need to plug the political leak in the Constitution if they want to rest assure democracy and liberty to be with them forever. The only way to plug this leak is to write in the Constitution that this country will relentlessly defend the marketing rule of supply and demand up to its highest genuineness in this country, and that maintaining a positive and rejuvenating social reservoir is an imperative way to guarantee the nation's future. With this writing, while not allowing capital monopoly in this country, capital accumulation will not have a single reason to be demonized but every reason to be glorified. *Capital accumulation for the nation will be one of the highest and immovable doctrines in this country besides the Christianity pillar.* Americans: declare this doctrine shamelessly with full bravery; absolutely no shyness is needed. Anything hurting this doctrine must pay a cost when trying to sham into the nation's political operation but must also pay a cost to face defeat. Without this new (but actually old) doctrine in the Constitution to guarantee the nation's future, America will have no future. A nation whose citizens think of no future deserves no future. This is a plain political mathematics.

To introduce a new clause in the Constitution takes a majority from the citizens for its approval. Doing this, the majority is to introduce a long-term principle in the nation's highest law document. We call this majority a long-term majority, in contrasting to the short-term majority that we will discuss later. With the setup of the long-term majority, people do not have to kneel at short-term majority so easily any more, but have a stronger backbone to insist what is right for the long-term benefit of the nation.

"See how ugly capitalism is, they even need to put monetary terminology in their constitution!" scream the socialists, grasping a "good example" to educate the masses whom they want so much to be their followers. Dear Mr. Socialist, if you do not want money for yourself, why do you need to build a government in which you can lead, actually rule over and seize everything of the nation, and why do all your comrades must stay in power forever once they built such a government? Give one exception to your followers to prove the opposite ever since the Communist movement starts. If you think monetary terminology being so ugly, why must you use the same term to attract and recruit follower?

Some of the modern socialists even tried to corner some American president to openly falsify capitalism in the UN meeting. For that purpose, they do not hesitate to display how capitalism has inexcusably victimized them with all the following criminal products: the gorgeous dress they are wearing; the sound system through which they denounce capitalism; the airplane that has taken them to the soil of America; the magnificent UN building in which they can stain the floor with their saliva; the camera in front of which they can exhibit to the world how their faces can display justice and anger. One can safely bet that when these noble and elegant socialists get back to their countries, they will not go home donkey back riding, but in a nice-looking air-conditioned automobile, another capitalist criminal product. Does one need to wonder if the toilet in their home to be a pit dug in the bush or something equipped with a flushing tank or valve, one more capitalist criminal product? Oh, people, shut up! Don't you ever try to speculate if their daughter has a Barbie Doll set or their son has a Nintendo game set! Your guess is so right that their children, ever since their birth, have only visited witches and sorcerers when getting sick, but never any doctor trained from the capitalist campus. So it is not the juicy portion of the capitalism that the socialists hate; what they really hate is the portion of capitalism that they have so far not found to be put under their control. To prove they are not hypocritical in hating capitalism, no better thing can be done by them to restore all the rain forest that they have chopped down in exchange for the refrigerator now in their kitchen. Then, after the restoration, let them show to their followers how they will chase after their daily food with a spear or a club in the forest while running bare feet. It is so much better off for everyone in the world if these socialists admit that it is not the capitalism that hurts them; what hurts them is their unrealistic appetite and premature greed on the capitalist juice coupled with their incapability of installing one healthy capitalist system on their soil.

Whether or not a society is genuinely democratic should be measured by the following standards:

a.   Has it provided an environment for the rule of supply and demand to be carried out with the highest genuineness? Or has it increasingly impeded the same rule, whatever the reason under which the impedance is proposed?

b. Therefore, has it provided the best operating mechanism to maintain and rejuvenate the social reservoir with the least stagnancy? In order to safeguard the social reservoir of the people and for the people, the only sound operating mechanism is that the social power of check and balance is left in the hands of the *majority* of the people.

c. However, has the power of check and balance been limited only within the sector of right, but fundamentally ignored or even abandoned between social obligation and right? Negligence of the check and balance between obligation and right, democracy must sooner or later slip into excessive democracy. Complete destruction of a social reservoir is the end of the society, whatever the society it is; ceaseless destruction of a democratic reservoir led by excessive democracy must lead to the end of the democratic society.

To guarantee the democracy, majority rule in democratic society has long been employed as one of the vitally important method in enforcing the will of one group of people over another group. While seemingly no better way can be found as alternative in will enforcing, if violence is to be avoided, democratic society has gotten used to a misconception that the will of the majority must be the absolute arbitrator, regardless the situation. Under this misconception, such idea gradually floods: the majority may not be correct, but doing what the majority wants must be correct. Unfortunately, being with majority or minority, every person must be governed by human instinct, which must lead everyone to prefer "right" to obligation. In other words, greed must more or less, intentionally or unintentionally, step in to influence or even grasp the minds of people, who in many situations may have grouped as majority in the society. The grouping may not be permanent but only last from case to case with different social constituents at different time for different situation and different demand, but such temporary grouping may be bad enough to be taken advantage by factions of special interest. This temporary grouping of majority, which we call short-term majority, is an excellent strategy for such factions to launch a revolution without the need of bloody social clash. All what those factions need are patience, distortion of concept, and eloquence, which includes sympathy begging, threat, and all kinds of "for your own good." Calling on bloody clash may not enable them to find enough followers, but that "doing what the majority wants must be correct" may; herded by human instinct, what the majority wants most of the time is "right," but lot less frequently the principle. One can hardly succeed in convincing a stranger to follow him to rob the bank. The same one may be very successful in convincing a big gang of strangers to follow him to "share" the national wealth. Both ways of getting access to wealth have skipped the procedure of obligation fulfillment, and therefore fit no better to the essential nature of robbing. However, the second way gives many people the illusion that "doing what the majority wants must be correct."

The aforementioned factions may not have a consistent and centralized leading core, a complete and "perfect" constitution or charter to guide them to do what they are doing. However, doing one step at a time, procuring profit one at a time, inheriting the same set of mind, each newer generation of such factions just take advantage of what have been corroded away by their forerunners. If the populace in the democratic society continues the blind superstition on majority rule, all these factions will eventually get what they want: dominance over a social reservoir. To make it more threatening to the society, in many cases, these social factions are actually minority in the society on the agenda they push forward. However, through eloquent convincing, they can successfully soften the resistance of some people. Then, this softened group is abducted as a group standing on their side, giving the society an illusion that they are sided with majority. An obvious example is the homosexual issue. It is a commonly well-known fact that homosexual people cannot be the majority in the society, but their agenda seems so invincible as if they were the majority in the society all the time.

Blindly believing in majority rule but disregarding how the majority has been temporarily formed for each individual case, the society gradually "learns" more and more to stay away from principle. This in turn enables the minority to pick up more population who has already been brainwashed to worship a new set of value. In 1950s, American Communists may have advocated revolution aiming at overturning the American government, but their slogan did not exhibit an openly robbing nature such as this: "We will not stop until we share all the national wealth." Today, in a society that a new set of values has been approved in the mind of a big population, the same slogan is held high by some widely spread organization in America. A big population's innate logic has been assimilated by and oriented with such robbing nature. In case someday they were able to realize their slogan through majority rule, should the society really believe that democracy has reached its highest genuineness? All these is to say that majority rule can be an important way to maintain democracy, but it can also be a way to degenerate democracy, depending on how the majority has been consolidated and ideologically baptized.

A tenant may be the same person who opposes the increase of property tax but demands rent control, as both the opposing and demanding are in favor of his interest: less rent. By opposing the increase of property tax, he would side with property owners at the ballot box; by demanding rent control, the same person waives a baseball bat toward the property owners at the ballot box. The same person joins the different group to form a different majority at different time at different situation for different issue, but all joining is motivated by short-term interest. Therefore, majority rule is not a dependably imperative way to stabilize the social structure in democratic society. The aforementioned factions can take advantage of the same tenant's will to form a short-term majority at different issue but then to serve their long-term goal. By successfully objecting the increase of property tax, they successfully restricted the government's ability of social management (let's assume the government has not been corrupted); by successfully passing the rent control, they successfully broaden their basic social support, both on mass population and injection of a new set of social value in people's mind. Their long-term goal, which is to dominate the society from bottom up, is realized one step at a time. Of course, this long-term goal is not their ultimate goal; their ultimate goal is to usurp the government power at the lowest cost, but certainly at the expense of the maximal willingness of the populace. Each minority group can make a political deal with other minority groups to form a temporary majority alliance in realizing each of their own interest.

There is no need to put up any example to illustrate how the society has been dominated from bottom up in the American society, the pus has swelled over the brim; we need only analysis on one weapon the aforementioned factions customarily utilize in forming short-term majority: discrimination. For anything an interest minority cannot have in any other way, he declares that discrimination has been engineered in against him, urging the sympathy from others who claim having been discriminated in other issues. One new majority is thus formed. People who have been so well "trained" in accepting guilt are so scared of such accusation that concession is easily made to keep the accuser happy in order to achieve a short-term peace. When the tentacles of these factions have reached the upper level of government, discrimination is easily made a persecution tool, and this further scares or even paralyzes people.

* A person demands the right to drive but insists the society to ignore whether or not he has auto insurance. If the society insists him to show his possession of an insurance policy before allowing him to drive, he declares that he, as a poor, is discriminated by the rich. The "rich," who has been guilt tripped for so long for "blood sucking" in history, easily let go of him. Actually, a big number of the "rich" is not rich but just be more responsible in the society and may even be as poor as the recipient of the "discrimination." Well, an accident happens. He who has insurance is held accountable, and the victim of the incident receives compensation to a certain extent. He who has no insurance, neither owning valuable belonging, shrugs at the victim. "Sorry, but do you think I want this to happen?" with both hands in pocket while softly whistling, the "poor" driver, or even killer, walks away from the victim almost worry free.

However, in case the "poor" guy is a victim of the accident, he will joyfully collect every penny from the responsible one. Why the society has allowed such mentality that, in this driving insurance issue, people with sense of responsibility has not been discriminated by the "poor" to the maximal cruelty? To make it worse, the community receiving the genuine discrimination has absolute majority in the society.

* A beautiful and brilliant young lady was blocked from accessing the title of Miss California because she expressed some idea that the homosexual community feels uncomfortable and subsequently, of course, discriminated. The point here is not to judge whose point of view has been correct or incorrect between this young lady and the homosexual community; the point here is who has been discriminated. As far as sexual preference is concerned, the masses can only group themselves into two groups: heterosexual and homosexual; the bisexual is actually a variety of the homosexual because the genuine heterosexual group will not be able to practice it. What if the judge of the beauty pageant happens to be someone who has heterosexual preference; he feels an expression from a homosexual candidate being offensive and blocks this candidate from accessing a certain title? With today's "standard," this judge must be fired because he is found discriminating the homosexual community. In a society that only two polarities are found, why has the society allowed a mentality that political pressure from one direction is a discrimination and punishable but the other direction is not? In accordance with such mentality, logic makes it so obvious that it is the heterosexual community that receives discrimination from someone's "anti" discrimination standard. To make it worse, the community receiving the genuine discrimination has absolute majority in the society.

* Someone desires to get into a certain school but unable to pass a certain standard. Instead of reviewing how much effort and willingness he has put in to catch up with other competitors before the standard examination, he accuses of discrimination from someone. Surprisingly, or not surprised at all, such guilt tripping strategy works. Gradually, more and more schools either remove the standard or lower the standard or even reverse the standard. The same mentality gradually infiltrates to other fields, such as English test in naturalization, passing certain examinations to get a government job. For example, in a school entrance examination that 100 points are full credit, instead of setting a threshold point of 75 for everyone, the school sets 85 as a passing point for someone who is assumed having been from a more socially "favored" group, but 60 for a "less" favored group. In some extreme case,

someone from the "less" favored group does not even have to pass a certain point because vacancies have been reserved for them regardless how many qualified ones from the socially "favored" group are purposely left out. Why has the society allowed a mentality that discrimination inevitably produced from the "anti" discrimination is not discrimination? To make it worse, the community receiving the genuine discrimination has absolute majority in the society.

No more is needed to illustrate how "anti" discrimination has been persecuting the majority in the society with ever escalating scale. All what people encounter in the nowadays American society is all sorts of "anti" discrimination. In too many cases, even the majority rule fails, but the minority with a flag of "anti" discrimination prevails. When the minority succeeds in all fronts with their agenda, the society would have accepted a totally new set of values, but up side down values, awaiting some "messiahs" coming for their rescue. Discrimination has become a phobia that nobody dares to have anything to do with it, not knowing that so many discriminations are necessary in order to maintain a normal society.

* A piece of merchandise must discriminate a person who has insufficient fund for the purchase, although the salesman may have to have the same smile toward both the "richer" and the "poorer."
* The society must discriminate someone from having the right to drive on the public land if he fails to provide auto insurance.
* The society must discriminate someone from having the right to drive if he fails to demonstrate possessing proper eyesight.
* A mountain-climbing team must discriminate someone who has no mobility without a wheelchair.
* The military must discriminate a woman from taking part in a big sector where arduous physical strength and swift action must be demanded.
* A Super Bowl Ring must discriminate those who fail in the competition from having it.
* A nation must be inevitably discriminated up to being wiped out by other countries in case such a nation is overwhelmingly homosexually dominated, cutting off the supply of successors of the nation. In order to avoid this from happening, what should have been discriminated or not allowed to be discriminated before the tragedy end approaches? Think, intelligent and responsible people!
* In making loans, bankers must discriminate those who show no ability to make repayment. Not allowing discrimination on this matter is a governmental guarantee of "lawful" robbing.

* In marriage, a person has full right to discriminate against anyone in all aspects including gender, behavior, and various backgrounds such as race, ethnic, culture, property possession, religion, nationality.
* A superstore owner must have full right to discriminate a Muslim in employment in case handling of pork is included in his business, unless the Muslim promises with unconditional willingness to take part in such handling when needed, doing the same thing like any other employee under this owner. The store owner does not even have to warn him beforehand; the store owner absolutely has full right to determine when to start or stop a pork-handling department in his business.
* A public library must discriminate someone from entering if such person radiates intolerable smell.
* A restaurant must be discriminated from continuing its business if it does not meet a certain sanitary standard. The health department has full right to concentrate its inspection on some particular kind of restaurant if statistic shows that some particular kind of restaurants has brought up more problem than the others.

The list can continue, but if "anti" discrimination prevails in all the items in the above list, can we tell how many individuals in the majority would have been damaged? The right to maintain discrimination is one of the rights for a citizen or an organization to maintain liberty and should be fully protected by law, so long as the liberty thus produced will not generate harm to the society but otherwise will. A Boy Scout troop should have full right to discriminate a girl from joining; by the same token, a Girl Scout troop should have full right to discriminate a boy from joining. To discriminate is to maintain the freedom and harmony of each organization, to remove the discrimination must cause harm to each of them. Of course, allowing either gender to form the scout troop but forbidding the other is a discrimination causing harm to the society, and the liberty produced by such discrimination must be eliminated by law.

Nevertheless, the existence of legitimate discrimination does violate someone's interest; or "better" yet, to disable the legitimate discrimination does enable someone to gain interest. Indeed, the "nobility" of excessive democracy does enable someone to gain huge interest in the process of removing the legitimate discrimination. To remove it, the "heroes" confuse the legitimate ones with all names of discrimination from other fields. So do flood in the society "discrimination against certain race," "discrimination against gender," "discrimination against sexual preference," "discrimination against handicap," "discrimination against poor," "discrimination against age," "discrimination against body size." Almost any legitimate discrimination must end up being deadly constricted once the heroes coil it with one or two of these flooded discriminations. Do not forget, almost all of these flooded discriminations have been made certain to be one-way weapons too. Anti "religious discrimination" will make sure that non-Christians have been found discriminated and such discrimination is punishable, but never that Christians have been found discriminated although the discrimination is more than obvious, let alone whether or not the one who practices such discrimination will be punished. In the Duke Lacrosse case, if the victims were three black boys and the woman making false allegation of being raped were a white, news of no less than racial persecution for a purpose must have worn off all microphones of international network system in proving how inhuman the United States has historically been. Upon the time that the three boys acquit and regain the freedom, the DA and the white woman must have all been politically crucified. Sorry, boys, you are white, even "racial discrimination" has been devised not in defense of you, let alone racial persecution. Someone has made certain that you can only learn to lick the wound at a quiet corner but with a blank memory about your nightmare.

All the above is to say that factions with Socialist nature have been successfully forming temporary majority in many individual short-term issues with various weapons and interest combinations to erode the society. One cavity at a time, a lion's canine will eventually fall off. People with higher intelligence must organize a rational majority to write a long-term principle in their highest law body, the Constitution, under the nation's birth certificate. With this principle, people will tell the world not only how this country is structured, but also tell the world what principle this country stands for. Democracy and liberty are too abstract as principle to be specified, but no rigid scale can serve its measurement. On the other hand, capital accumulation and free marketing rule of supply and demand can be accurate down to each penny as material mean to guarantee democracy and liberty. With this set of principles that are backed by long-term majority, all short-term majorities can be put under control. With this set of principles, people can prevent the socialist tentacle from reaching and grasping our Constitution for their service.

Chapter 19

In Need of a Long-Absent Institution

The multiparty system in the democratic society has one advantage: the check and balance on power. However, the disadvantage of such system quickly shows up to outweigh the advantage. The disadvantage is its selling behavior that is forced out of each politician. In order to survive in the political arena, competition between them forces them to promise all kinds of rights to the masses in order to get the most counts from the ballot boxes. Each election campaign is virtually a fierce rivalry of promising of right. Aren't the Western people too familiar with this? However, obviously, anything the politicians promise to give is not from their personal treasure vault but from the social reservoir. Some of the promises even go so overboard that the rights they promise are not what the nation can afford at all. It does not matter; what matters is to have the competitor overcome. Gradually, this mode of competition advances the reservoir outlet of right to a gate, then to a floodgate, and then to an irreparable floodgate. The first victim, but also the least noticeable victim, is the national reservoir. In a nondemocratic society, those who can benefit from the depletion of the national reservoir are only few, but in the democratic society, the number of those who join to deplete the national reservoir must be tremendously out of proportion. The national reservoir quickly suffers.

If one party can be so successful in competition of abusive right promising that she can usurp the absolute power of a nation, then the political price tag from that party will change in another way: no more right promise from the party. Instead, only a rosy picture is promised by the party, but the populace is made to promise away their rights. Changing the presidential term from a four-year term to a twenty-five-year term in some country's Constitution is only step one for the populace to promise away their rights. This monoparty behavior has been proven by numerous historical events. To prevent such monoparty situation from happening in democratic society, people must first stop the election from becoming a rivalry of abusive promising of right. However, where do people get this power? Do not forget, in the case of America, all power the political candidates seek is the power bestowed by the Constitution; and the power that enables them for the seeking is also from the Constitution. In other words, their power cannot go beyond the Constitution. There is one utmost supreme document above the Constitution: the Declaration of Independence of America, the nation's birth certificate. Through the nation's birth certificate, the Founding Fathers obliged all inalienable rights to be countermeasured with mutual pledge. Such countermeasurement, both as obligation and right, has long been vested in the American people but so far is set aside. Therefore, the American people are obliged as well as have the right to demand the establishment of an institution to carry on the function of countermeasurement for the nation. Any right that is to be promised and granted to the civil individuals by office of any government branch must not resist to be weighted with the mutual pledge requirement by this institution.

The legitimacy for the institution in argument to come to reality does not stop at the concept of obligation and right but also from the point of view of the balance of a valid social contract. If a Constitution was drawn to secure the inalienable rights found in the nation's birth certificate, an instrument to guarantee the mutual pledge requirement found in the same document is certainly equally required to intervene the national affairs. When it is time to detect its absence, it is time to have it installed. The entirety of the nation must consider its establishment no less important than the establishment of the Constitution. From the point of view of Mother Nature's golden rule that obligation input must be bigger than the output of right in maintaining a social reservoir, to set up such institution to secure mutual pledge between fellow citizens is even more vital to the nation than any document to guarantee right. Indeed, its absence has long infringed the genuineness of the function of check and balance in the democratic society. Since such an institution is demanded and supported and empowered by the national birth certificate, its establishment is then independent from the Constitution. To have this institution established, no voting is ever needed, and no one should have power to stop it from being established. Did the populace vote to empower the Founding Fathers to bring the nation's birth certificate into reality and launch the American War of Independence?

That the U.S. Constitution should find no legitimacy in establishing such required institution is obvious, because the Constitution is subordinate to the nation's birth certificate. Allowing an institution established by the Constitution to check how the function of the nation's birth certificate has been carried out is like allowing a son to supervise the parents. It is offensive. Besides, the U.S. Constitution is basically a social contract promising right to the citizens; in many areas, it just cruises at the edge of being a fake contract, allowing too much illusion for people with special ambition to flex. In some situation, it cannot even find itself powerful enough to maintain the rigidity of its words. Let's look at the following example:

"Neither the United States nor any State shall assume or pay any debt or obligation incurred in aid of insurrection or rebellion against the United States, or any claim for the loss or emancipation of any slave; but all such debts, obligations and claims shall be held illegal and void"—this is a clause one can find in the Fourteenth Amendment, adopted on July 9, 1868. This clearly tells people who the emancipator is and how powerful this emancipator can be. Rigid as these words have shown in withstanding the claim from the previous slave owners, however, they are seen lacking enough rigidity to withstand the challenge from the ones whose ancestors are liberated by this emancipator. Needless to say, the previous slave owners must be seen as in "rebellion against the United States" in this amendment. With respect to that time, any political pressure against the United States at that time must be seen as an "obligation incurred in aid of insurrection or rebellion against the United States." As the extension of the Constitution, this amendment thus declared "all such debts, obligations and claims shall be held illegal and void." While no regime change has been found in the United States, while the Constitution's words remain the same, a debt or obligation against the United States imposed 150 years ago was "in aid of insurrection or rebellion against the United States," but somehow political pressure of the same nature imposed by modern citizens is not seen as so negative anymore. Believe or not, the Constitution has shown signs of losing rigidity about "illegal and void" toward the modern "debts, obligations and claims" of the same nature against the United States.

Neither does the Constitution have reason to stop the institution supervising mutual pledge from being established. "The powers not delegated to the United States by the Constitution, nor prohibited to the States, are reserved to the States respectively, or to the people." This is the Tenth amendment of the Constitution. When the American people feel obliged by the Constitution's mother document and demand to exercise the right bestowed by such document, the Tenth Amendment clearly expresses the position that the Constitution should take. Such institution accepts no check and balance from anyone, but the term "we" found in the declaration can have all arbitrating power in completing its check and balance. Anyone intending to stop such institution from appearing must convince people why power defending the very first document of the nation at her satisfaction must be made absent. She demands this long neglected power to be resumed now; she feels that all powers produced by all documents subordinated to her have left space to be taken advantage by attempts aiming at discomposing her.

The American Supreme Court cannot stop such institution from appearing. The existence of the Supreme Court is because of the Constitution. Her power has been historically found not going beyond determining an issue being constitutional or not. The court verdict regarding the Ten Commandments and Alabama State building fully exposed the court's limit: the principle that the court applies regarding the conformity to Constitution on a certain issue fails to determine whether or not the words "Creator" and "Divine Providence" should be removed from the nation's birth certificate.

The American president is also seen as a child to the nation's birth certificate. He is the number one citizen to be well protected by the nation, and naturally, he is the number one citizen to be required by the Mother to mutually pledge to the fellow citizens of the country. A pious child can only devote all his intelligence and ability at her service other than stop her from realizing something.

Possibly, the most natural establishment to push forward the appearance of such institution is the Congress, because her political formation has the most resemblance to the assembly that the Founding Fathers put together when they signed the Declaration of Independence. The Founding Fathers' assembly was the political body that directed all the nation's affairs during her infancy year before all the government branches brought up by the Constitution came into reality. On this matter, the Congress should play a role of a midwife other than an abortion doctor. If the Congress happens to work like an abortion doctor, she should present to the American people with a document in which her power serving such purpose had been so invested.

When the institution of supervising the granting of right has any chance to come into reality, it must faithfully reproduce one vital character shown by the assembly of the Founding Fathers: They did not declare that they singed the document as a representative member from any party but as a representative holding the sole belief of Christianity to sign for the people. Therefore, it is only natural that anyone empowered to form the institution must be people from no political party. To guarantee this, he/she must take oath to have himself/herself isolated from any political party; if he/she ever has membership with any party, the oath thus forces him/her to have party membership relinquished. In order to guarantee to maximally eliminate party influence, a law must be made to forbid them for a certain period of years from joining any future activity of a party, not even fund-raising or giving a speech on behalf of any party.

Based on the way the Declaration of Independence is written, the people who can most genuinely continue the Founding Fathers' will should be people holding the belief of Christianity. Only Christians will perceptually conceive the Creator and the Divine Providence in the same nature as the Founding Fathers did, unless people can prove that the Founding Fathers are the infidels to what they believed, and therefore, the Creator and Divine Providence can be from somewhere else. So, when the institution that can faithfully function as the birth certificate requires, only Christians can be the member of such institution.

However, make no mistake; to establish such an institution is not to empower anyone to add power to or to subtract power from any government branch. Nevertheless, the Declaration of Independence itself is not a detailed document about government construction, although it is the only indisputable document to enable the existence of the nation and, subsequently, her government. The only function of the long-absent institution is to monitor how a right that a citizen receives has been genuinely and faithfully measured by the mutual pledge that he/she is supposed to put up. Her function thus will not go beyond civil level.

Besides being a Christian and neutral to any political party, each member should also include the following personal qualities:

1. He/she must be an American citizen, either naturally born or one who has been naturalized.
2. He/she must take oath to confirm holding the belief of Christianity and has had proven record of uninterrupted activity in an American church for some minimum years, such as five. The spectrum of Christianity may cover members from Roman Catholic, Orthodox, Protestants of all denominations, and Mormons, and others who all revere Jesus Christ. However, any organization that practices human sacrifice ritual or money extortion in any manner in the name of Jesus Christ must be barred.
3. He/she must be a parent who has at least one offspring and is still currently retaining the only wedded spouse to prove his/her integrity in holding the seriousness of pledging.
4. He/she must have minimum of two years of college education with a diploma; a dropout cannot qualify. This education requirement is based on the fact that the Founding Fathers are all comparatively well educated in their time. The college issuing such diploma must be a college with approved credential from the American government, has her major campus on American land, and uses English as her only language in all lectures except in classes teaching foreign language.
5. When the country has a chance to set up a hierarchy system of citizenship classes, each candidate must be a first-class citizen—no excuse.
6. The membership of this institution will not link to property ownership (but no bankruptcy), gender, race, geographical origins, although minimum years of membership of Christian church of American origin cannot be compromised.
7. Each of them serves the institution only one time of one—or two-year term in his/her lifetime; in some way, they work in a similar nature like a citizen serving the jury duty.
8. The members will not have a full-time office but only be summoned by a full-time secretary of this institution when needed; the assembly

will be minimum once a year for a duration of two or three weeks or as frequently as the secretary determines.

9. Although summoned by the secretary, they are not subordinated to the secretary.

10. The job of the members is to review a list about rights that have been funneled to the secretary through various channels. They will decline a right that they considered as having been improperly granted or approve the granting of a right, which has been granted to some individuals by some office but is under protest by someone else. For example, when a murderer has been granted the right to extend his life by a judge, but the family of the victim protest, the institution can decline such right to be granted to the murderer or approve the granting if they see fit. Or when a spy is charged of a serious crime that only death penalty is justified, but for some reason he escapes such penalty, members of this institution can recapture this spy with death penalty if someone, such as a law enforcement agent, makes a complaint against his escape.

11. This institution can only show consent about right that has been or shall have been granted by some government offices, but members of this institution cannot create right and then grant. The job of this institution is fundamentally "not to grant" but not "to grant." In the death penalty cases mentioned above, they cannot overturn the judge's decision in punishing someone with death penalty.

12. When the institution considers consent of granting as being more appropriate than declining, two-thirds majority is required. For any right they decline to grant, simple majority is needed. Do not play "fair" and "unfair" concept here only because of the different requirement of majority. At the time this book is written, report says that each American household has on her back a national debt of $560,000. How fair has this generation been to the generation that must come but not yet?

13. One member is selected from each state. So there will be 51 members including someone from Washington DC in the new institution. Such member number can be expanded by a multiple number such as three or five or even seven across the board to every state.

14. If for any reason a state cannot have enough members to join the institution, this will put the state in a position to decline a right in controversy. A member who expresses in any language other than English during the time of assembly automatically disqualifies himself/herself from voting in the institution, regardless.

15. Absolutely no lobbying of any kind is allowed to approach the institution members except flyers or booklets displayed in open space for everyone's reading.

16. Each state should always have a list of the churches that are willing to provide such candidates for service when needed. Each church must offer record showing minimum constant church participators. Churches with fewer numbers of participators can unite to produce such a minimum number, but the union must be registered before the selection time, say, two years.

17. With the popularity of Christianity found in America, possibly, it is good enough to serve the purpose that only churches in a district for election of state representative are needed for each selection of members for that state to join the institution. Next time, rotation will make it another district to be summoned. No district will be repeated until selection has been made in all districts in that state.

18. Possibly, racial problem is the most difficult one to deal with. How about this: Candidates will be selected from the four most popular races in the district plus American Indians, regardless what population the Indians have at the time of selection. American Indians should always have some natural privileges. Any race that is considered popular in a district must have ten percent or more in the population. American Indians will be counted as minimum of ten percent but can be more if they actually are. Calculation of the frequency of candidacy for each race to join will follow the example that is illustrated in the following. In a district, white is found sixty-three percent among the population; black, fourteen; Hispanic, sixteen; Asian, five; and the rest are the Indians. A race that has ten percent or more among the district population will have the percentage figure truncated to a single digit. So with the above figure, white should have a digit of six; black, one; Hispanic, two; and Indian, one. Add all these digits together, we get 10. At each time of selection, a member is chosen from the race that has the highest number. However, each time a number for a race is used, the number becomes one lower. When the selection is going to be done in the next district, the same race will have a lower number because of the selection done in a previous district. Each time the members are produced, they should be all from the same race, be the multiple number three or five or seven. When the cycle number of ten is used, a new cycle starts with the next district in turn. Each time the selection is moving to the next district, the racial composition may be different in the upcoming district—so is the cycle number—but the frequency number that a race has used must be applied to deduct its potential to join the next institution. A genetic mix should be allowed to identify himself/herself with a race from either parent for the purpose of joining the institution, but such identification should be filed in the court and will never be allowed to change again for that person.

19. Being a member of the institution is a public service of no payment. The enjoyment for them is the honor of guarding the nation's energy reservoir and a government-paid trip to the capital with room and board plus full compensation of loss of payment due to absence from job. At most, a "tip" is awarded for the time they are away from the family. A memorable plaque should be given to every member at the end of their term to thank them for their service to the nation. Each member should be exempted from local jury duty for some time. With this condition, the competition for becoming the institution member should not be fierce. If no volunteer is found, someone can be summoned in a way similar to serving jury. However, an individual can have full right to decline the summoning call on the reason of retaining his/her membership to a certain party.

20. Since the institution is not meeting on a daily basis, the institution should thus have an office with its own full-time secretary of full pay to supervise the daily business of the institution. The responsibility of the secretary includes summoning meeting of the institution, collecting and listing protests from all source about abusive right granting, presenting such list to the members, explaining why a prospective right granting is seen as abusive, and supervising the voting of the institution members. The secretary must constantly review promise made by candidates running for government office through election and forewarn the people which promise may be potentially invalidated by the institution. This job is particularly important in the eve of presidential campaign. The secretary will have no right to vote in the institution but just count the vote. No institution member is subordinated to him/her. His/her job should, through proper representative agent, also include supervising oath taking that involves the clause "We mutually pledge to each other our Lives, our Fortunes and our sacred Honor."

21. The secretary must be able to remind some officers of their negligence of duty due to their selective law application between different groups of people but for similar situations; he/she should also plead the action from the Congress for correction of such selective application. Law is only as good as it is applied. The political outcome of the same case can be completely different whether law is applied or not. Law will lose its fairness in democratic society if it is allowed to be manipulated.

22. The office of the secretary must be equipped with a propaganda team that is fully facilitated with all kinds of advocating medias such as TV, radio network, newspaper, and Internet. The function of this team is to incessantly emphasize the importance of mutual pledge in building an ever-strong nation; to counterbalance the propaganda of right whining;

to explain how impractical it is for some promise made by some political candidates in their attempt to win election; and to honor good citizens, particularly those from public service, such as the military soldiers, policemen, firefighters, and teachers (public school or private school). Propaganda is one of the extremely successful tools for the Communists to realize their agenda, and history has proven that their propaganda is only lies after lies. However, so many people continue to believe their lies and feel so drunk. It is only stupid for capitalism to give up such tool and let the Communists play this tool in any way they want. If capitalism continues to ignore the importance of this tool, she is no more than willing to accept ultimate defeat by her enemy, who has so much excellence in propaganda of lying.

23. The secretary also has one extremely important job: He shall make sure that all schools, public or private, regardless religious sponsoring background, from primary to high school, provide courses to educate students their national obligation, social obligation, minimum citizen conduct, and the importance of mutual pledge as required by the Declaration of Independence. Education based on racial hatred and class hatred must be corrected and forbidden. While mentioning the historical pitfalls of this nation is unavoidable and should not be avoided anyway, the educational system must emphasize how honorably this nation has had these pitfalls corrected. Education that paints this country with sin and guilt must stop; what the students should know is that, in human history, no country like America is found so willingly to spend so much effort and sacrifice for self-correction. The reason that the student can be on this land is because of the glory that this nation has created for every citizen but not because of the crime that someone commits. Education based on hatred against this nation is not a freedom of speech expressed by the teachers but a crime of brainwashing that serves the interest that the enemy of this nation cannot wait to harvest. The teachers' "freedom of expression" in preaching hatred to this nation makes the enemy's harvest so effortless; such "freedom" must be a punishable crime.

On the other hand, however, while preaching hatred against this nation in the classroom should be punishable, the institution should recommend to the Congress to protect the teachers' authority to some extent. Some of the teachers' teaching manner and teaching method must be immune of lawsuit unless it constitutes an actual crime such as child molestation. In the list of the abusive right pursuers, teachers' authority is one of their targets too.

24. To avoid party abduction, the job of such secretary must be reserved for a person who is not from a party that holds majority in Congress. He is not necessarily required to be isolated from any other party. The majority of Congress is determined by the total sum of members from both houses. While in office, the secretary will not run for any government office through election campaign. If he decides to run, he should submit resignation within a certain time frame and wait for a replacement.

25. The decision made by this institution should face no overriding from anyone because the power of this office is produced by the most primitive but the grandest and most honorable document of the nation. Indeed, this institution should have power even to review rights that has been granted "too long ago."

26. Any member can have the right to demand the right under discussion to be precise. Otherwise, a broad term like "human right" will only bring disaster to the nation if approved.

27. The oath for each member to take for assuming duty should include "And for the support of this Declaration, with a firm reliance on the protection of divine Providence, we mutually pledge to each other our Lives, our Fortunes and our sacred Honor."

28. No member in this institution is immune from law.

29. Besides reviewing the matching between a civil right that is to be granted by some office and the mutual pledge according to the nation's birth certificate, this institution should also dutifully recommend to proper government branch the termination of a certain officer who is found either having abused his office power in granting right to civil individuals or showing pattern of selective application of law in favor of some particular group. The termination may be either through executive order at a proper level or impeachment. So officers who enable the American prisoners to escape from involuntary servitude should be also in the institution's list of recommendation; words from Amendment 13 must be made as rigid as they can get.

30. Institution of similar function should also appear in each state. At the state level, the secretary for that state institution can have the power to block a proposition from entering the ballot at the recommendation of certain number of state assemblymen and/or state senators; majority is not required. The proposition to be blocked is one that has obvious Socialist nature, which will enable personal enjoyment to be converted to social burden while the one receives enjoyment is not obliged to contribute accordingly. The pass of such proposition only ends up enabling some political individual to gather power for social extortion.

31. All members must be aware that this institution is not a place for religious practicing but a place to check the balance between rights enjoyment and mutual pledge with the highest fidelity of faith in continuing the Founding Fathers' will. Besides citing "Help me, God," any suggestion of worship or Bible reading inside the institution may be found not matching the Founding Fathers' idea and should not be allowed.

While the Bible has condensed abundant wisdom in guiding personal conduct, it is hardly found as having condensed principles to guide national behavior or government administration. It will only lead to tragedy to formulate governmental behavior with ideas concluded from personal conduct. While love and forgiveness should receive admiration at the level of personal conduct, a nation must be buried at no time if guided by the principles that "if someone strikes you on the right cheek, turn to him the other also." All institution members' power is given by the nation's birth certificate, not by the Bible. Above all considerations, that "for the support of this Declaration . . . . we mutually pledge . . ." always takes the first priority. Every right in dispute must be measured by this consideration before showing consent for granting. Not holding the culprit accountable is mostly in direct violation of the mutual pledge that is required of every citizen by the Declaration of Independence.

In God's kingdom, the space reserved for personal conduct to display must be smaller than the space reserved for national behavior. It may be God's purpose to leave a bigger space for the people whom he loved to display their own wisdom. Therefore, government behavior and government administration are not apparently revealed in the Bible. Stupidity and naiveness shown in the smaller space may receive forgiveness or mercy from God; the same will be seen as irresponsible behaviors in the bigger space reserved for national behaviors. Irresponsibility must receive punishment. On the cross, Jesus prayed for those who were crucifying him, "Father, forgive them, they know not what they do." This is certainly sublime nobility at the level of personal conduct. However, in the space reserved for national behaviors, God asked his son to leave at a very young age. In the space where the national will of Rome needed to flex, with the principles that Jesus preached in the space of personal conduct, he felt that he must cried out, "My God, my God, why have you forsaken me?" At the personal level, God forgive humans' sin through the death of Christ. However, at the national level, the message through the blood of Christ is clear: let value of no one's life surpass that of Jesus's life in guaranteeing a nation's continuation.

All humans have simple mind, regardless how faithful and how devoting one may feel he is toward God, although someone can be found more intelligent than the other and therefore less bound by simple mind. With a simple mind, what one conceives as a way to please God may not be necessarily the way that God accepts as no alternative. When Abraham was about to kill his son to please God, God had an angel to tell him to stop. The way that Abraham pleased God was even conceived by him as a direct instruction from God. It would certainly displease God for someone in history to have perished so many of his children only because of a controversy over the significance of bread and wine. While no single ritual in worship has been revealed by God as the only style to please him, he certainly feels displeased if some of his children invoke difficulty onto some others only because of different style of ritual. His displease has been obvious to some race, and such displease is reflected by population pressure that this race is suffering. Is this also an apparent revelation from him that his children have been in a critical epoch? The revelation may be that searching for common ground between each other and unite will serve them tremendously better than searching for difference and divide. Are difference and division really so worth insisting compared to the threat that Christians must be facing in this country and in the world nowadays?

This author is no law expert at all. All the above discussion is only something that interests this author for some humble imagination, hoping that some ideas of genuine intelligence out there are attracted on the issue on how to better guard the national reservoir. The fundamental idea is to hope for the forthcoming of some priceless wisdom and work from someone that can contribute to keep the nation's reservoir perpetually vibrant. If it is finally proven that an institution to safeguard *mutual pledge* cannot be brought into existence, all it can be said is that some people objecting it have been made too powerful. It matches their special ambition if the part about mutually pledging from the nation's birth certificate can be made invisible. The document of grant deed is useless for a home owner unless sufficient power backing this document can prevail.

The grandeur and loftiness of the nation's birth certificate rested on the fact that, when signing, the signers must be aware that a document so written would eventually lead to threatening the interest of many of these men's future families. That "all men are created equal" was an open threat to the slave owner's interest. In spite of such threat, however, the slave owners signed it together with the others. The unique sensitivity possessed by politicians about value of words did not deter them from signing it.

What is clearly written in the document is "all men." If a racially slavery country was a social prototype in the mind of the majority of the signers, they would have done either of the following: (1) replace "all men" with "all white men" or (2) exclude the enslaved race from "all men." In the subsequent political practice, they would have forced their slaves to go to the bloody war front first. None of these is found in history. Instead, they signed the document, leaving political space for any man and all men, slave owners and slaves alike, to be equal when conditions mature. It is this political space provided by the nation's birth certificate empowered the government to launch and win a war of emancipation; the other one that abandoned this certificate was defeated and vanished. It has been a historical fact that this certificate brought into reality the abolishment of slavery practice on this land. Can anyone prove that the nonexistence of the Declaration of Independence would have resulted in a nonexistence of slavery system on this land? Does any descendant of the slaves regret the very existence of the nation's birth certificate that is so written? Sometimes someone is heard to devaluate the loftiness of the Founding Fathers because their minds and their actions are not found to have been directed by our modern political standards. In doing so, this person must answer this question: how would he/she surpass the Founding Fathers with more sublimation of integrity guided by modern political standards if he could ever live in the older era? Or, living as a civilian in today's society, has he/she been thinking and acting according to political standards that will only be found in 230 years later in the future?

Chapter 20

Capital Accumulation

It has not been emphasized enough that capital accumulation is the lifeline of capitalism. This lifeline must be restored as soon as possible. Besides trying to revive the manufacturing lines as many as possible, social savings is also critically indispensable. If the populace cannot get used to this habit overnight, a responsible government should show the way to start it. For this purpose, will the following suggestions be seen considerable?

1. Increase tax to curtail the habit of overspending among the masses; this will also help to forcefully curb the fund (actually capital) from profusely overflowing to the foreign production competitors.
2. However, the tax increased for this reason is not allowed to be spent by any government office for any reason, except reinvesting in creating manufacturing lines.
3. When a project aiming at building a manufacturing company is proposed, the government provides the fund from this new tax; but the project will be sponsored and operated by a private business, who may be a sole entity to have proposed or a sponsor that is recruited through bidding.
4. Any private business that is accepted to manage this company but funded by government must post bond. Two kinds of bonds are required: a bond that requires the private business's possession as collateral and another bond that requires personal property of all executive personnel above a certain level as collateral.
5. Any individual can also propose a company for the government to consider if adequate bonds can be posted by him.
6. Once a total sum to start the company is finalized, no more funding is allowed from the government for any reason, even if the company is seen failing. All personnel that commit to the success of the company must be held accountable. Here is where the bonds stand and what they stand for.

7. The company can recruit capital from the society. However, the government's investment is always the first one to be protected when financial difficulty appears in the company's operation.

8. By a certain day that is stipulated in the contract between the government and the private business, if the company shows accomplishment of certain revenue goal and makes money, the company will be sold in the open market. If the purchasing results in excess over all the total investment, government and public fund together, after inflation is deducted, the sponsor will be awarded by part of the excess. The rest will be pocketed by the government for future investment. Anybody can buy this business. But if the transaction results as a loss, the management company is not allowed to buy but be held accountable for compensation of the loss.

9. In all this kind of companies, no organized labor is allowed. If the management group must be forced to face the fierce competition in the open market, it is only fair that the laborers must also face competition to certain degree. Indeed, the fierce price competition between merchandise is forced to be there by the populace; if the labor that the populace sells to the society does not have to face the same, this is an obvious extortion. Besides, in private premises, the owner should have full right to determine who is allowed to be in.

10. The law of minimum wage cannot be encroached. Workers in such government-funded company should have higher priority to be considered in getting welfare if the minimum wage cannot help to catch up the family need. However, there is a time limit. Workers can only improve their pay through skill level advancement or quantities of unit product completion.

11. Except for establishing a manufacturing company that is production oriented, no any other project, even in the name of research or service or a toll-collecting bridge, can use the money from such new tax. If no such company can ever be proposed, let the money sit there idly but at least earn interest from bank savings. More specifically, if the end product from such proposed company is not something that can be put in a showcase without preserves in a store, forget it. On the other hand, anything that has strong export potential, encourage it.

12. When the manufacturing activity can stand on its feet on this land again, and the savings habit is seen growing, consider to stop this new tax.

13. From now on, at least for a period of time, for whatever new business set up that is production oriented and is not using the government fund, law must be established to fully protect such business from the interference from any organized labor, although minimum wage

cannot be compromised. If the business is fundamentally making products for export, it should be exempted from carrying workmanship compensation insurance. For any suffering due to accident in production, the government picks up the tab. However, of course, if the accident frequency is above reasonable rate that is acceptable by industrial standard, the company will be forced to pay its own cost for the insurance, even be retracted back the payment for a certain period of time in which this company has been exempted. Besides, workmanship compensation must be put under the hawk eyes of law so that any fraud aiming at extortion is made a high crime. High cost of workmanship compensation insurance has been hurting the business on this land tremendously.

A genuine democratic society cannot exist without capitalism; capitalism cannot exist without the lifeline of capital accumulation; such a rejuvenating lifeline cannot exist if monopoly of either capital or labor force is allowed to develop. So sense of competition in this country must be restored to meet the competition level of the world. A country that has policy to immunize ineptitude from facing pressure of competition is doing no good to anyone but just self-eliminating in the international competition world.

So a genuine democratic society must be a society in which no monopoly of anything on the production line can be allowed; *capital accumulation must be seen as an honorable and indispensable social responsibility*. However, no capital accumulation will amount to any significance if the gigantic evaporator of national wealth, the current abused welfare system, cannot have a fundamental reformation. This system has not only been abused by those declining to contribute to the society but is also well abused by those who financially do well. In order to take advantage, some of them just transfer their possession to someone they trust, such as children or siblings, but then continue their enjoyment of social aid as a poor chap. Here comes a true story.

Under a mutual agreement, a person worked for some employer for cash, not paychecks that were linked to social security number. Without record of his income in the government, he received social welfare. After a while, with the wealth he was able to build up, he wanted to buy a rental property. He asked a clergyman to help. The arrangement he proposed was like this: The property was bought under the name of the church; the church would collect the rent as his support to the church, but the church would release the ownership of the property back to him any day in the future when he felt necessary. He continued his "working" style, of course. Well, the day that he felt the need to take back the property came. Argument arose as the clergyman refused to release the ownership as promised. Bad words were spread against the church. The clergyman then told this person: "You better stop what you are doing, or I'll report how you cheat on the social welfare." Knowing that he may have to deal with the court and lose, this person had to silence himself to avoid the jail.

Welfare abusing does not stop with this kind of story. Many immigrants, as soon as they stepped on this land, with a green card in hand yet zero contribution on tax, they were able to manage to receive financial aid from the government for various reasons. Then, a big part of what they received was sent back to their "great motherlands." Some of the money was used for helping the family back home; some of the money was even directly sent as a donation to some public or governmental projects for them to earn the praise of "patriotism." Does Uncle Sam ever know that part of the political and economical pressure that he ever increasingly feels from the governments of those great motherlands were "purchased" with his own money? Uncle Sam's foolish charity just tossed away money for reason but against him. Having been mocked as a paper tiger for more than half a century, Uncle Sam is now seen over feeding some true tigers in too many ways. The problem is that Uncle Sam's generosity seldom brought him praise, but way more often than not, swearing, hatred, and scorn. During Katrina of 2005, people could tell apart two groups of victims from the TV screen. One group was way calmer than the other. The other group seemed that their overwhelming swearing, hatred, and scorn against the government were never enough to make up what they had suffered. It is surely a pleasant thing to learn if investigation can reveal that the calmer group had a much bigger percentage of people receiving welfare aid so that Uncle Sam can feel relief: "My money had been well spent." Oh, with all these kinds of stories, we have not even included those who sell food coupons for drugs.

If the social mode in choosing political leader cannot be transformed from right promising to a mode of promising the ability to supervise and reinvigorate the social reservoir, the welfare system must be continuously abused by some people and such abuse must also be ill exploited by the politicians. The democratic society must inject such mentality to her social members: personal satisfaction is absolutely a personal responsibility, not a social responsibility. Being poor is not an entitlement of recipient, but an obligation to improve oneself in competition although with government's help if necessary.

Chapter 21

A Second Revolution

To bring America back to what and how our Founding Fathers had expected this country should be, to bring back the pride that each honorable American citizen deserves in this world, nothing can be easily done. The job will face so much resistance that it may even need a term called a second revolution to fathom its greatness. All of the following must be part of the goal of the second revolution:

1. English must be reinstated to her absolute throne as official language. Her absolute monopolization to this throne was long established by the Founding Fathers through the three most authoritative documents on this land; only ignorance, loss of will, shortsighted interest, and incentive supported by betrayal purpose allow her to be dethroned.

2. All Christians in this country must unite to restore this nation as a Christendom country. With the pace of the ever-increasing political pressure that they have experienced today, if they continue sitting there feeling lunatic and idle and only wait for "fairness" to come on this matter, it would only be a matter of time before the political pressure is escalated to persecution. The end result of this waiting is that they lose the Christendom and that Americans lose their United States; not only America will repeat the ending history of Rome, but Christians also repeat the Roman era before Constantine. Jesus has cried for them, "My God, my God, why have you forsaken me?" They have no right to wait and anticipate God to send another one to cry the same for them again in the future.

3. All monopolies, capital or labor force, must be gradually dissolved. However, no radical practice should be allowed to work for this purpose; otherwise, the country only asks for disaster. Monopoly of labor force can be dissolved only through compensation in a progressive way, but they must be dissolved if lines of manufacturing and production are to reappear on this land with the vigorousness they deserve. If capital

accumulation is not given a chance to regain its breath, democracy can only stare in her open eyes as Marx's dictatorship of proletariats is on the way to monopolize everything.

4. Caucasians must enable themselves to organize on a common ground but above any partisan passion and ideology and then enjoy all the political rights, which they have been deprived of but any other race can enjoy. It is certainly true that an organization of Caucasians will not be welcomed by racists and even be opposed by some "noble" Caucasians. Without such an organization, they must eventually lose all political backbones and be an underdog in every political competition where consolidated forces always have upper hand. If the U.S. Constitution is to make sense, it must allow either everyone to organize with the same equivalent background or must allow no one to organize following the same standard. Violating this, the Constitution must have allowed someone to degrade itself into a tool of dictatorship for some group of citizens over another group of citizens.

5. Uncompromised competition in all fronts must be reintroduced on this land, in the same intensity of fierceness that is enforced onto us by other countries. The only thing we can do differently from the others is that we do not remove the weak individuals in the competition but help them to improve themselves to be better competitors. However, if any individual must insist on self-giving-up in the competition, this nation should let not such individual hinder her from moving forward.

6. Americans must reform her political system and mentality in selecting leaders, switching from a mode of competing right promising to a mode of competing leadership inspiring contribution to the society. A true leader must be the one who dares to call on people, "Ask not what your country can do for you; ask what you can do for your country." A people with genuine wisdom must be those who can distinguish this kind of leader from those crooks who will tell them, "Give me power. I'll show you the way to get your right." America can no longer afford more right promising. With the situation as is today in everybody's eyes, any more right promising must only be done on the basis of deficit of politics, economy, and morals. America—the once grandest, most vibrant, and most dignified lady in the world—has been overbled. A true leader in today's American society must be and can only be the one who dares to and is able to lead the people to affirm the authoritative status of the principle of capital accumulation for the nation.

7. All Americans, regardless of race or ethnic background, must adopt this concept with full passion: The land, named America, is *My Motherland*. To those who are native born, it has always been this way. To those who

are naturalized, this is the choice they make with absolute free will; they must have been deeply moved by this young stepmother's generosity in deciding to join this nation. Only if every citizen entwines one's fate with the land that they would call mother can all citizens secure themselves with a well equipped carrier in the world's political ocean where surging of angry waves is a norm. To view the same land as a hostel next to a goldmine for polyglots to search temporary employment or fortune, all these citizens can only prepare their descendents to degrade to political orphans in the future.

8.   All Americans must relentlessly defend and utilize the authority that their Founding Fathers set up: mutually pledge to each other. This authority must permanently remain inflexible, immovable and unchallengeable on this land.

The job of reviving America is formidable, difficult, and, in some situation, may even be dangerous. No job is easy in defying the looming of Pluto or Grim Reaper, but it must be done—and be done on time. Americans have not been left with luxury in making choices except two: to defeat or to be defeated. The first choice is tough and very demanding on bravery, intelligence, and endurance; but only this choice will afford them a chance to look at a man in front of a mirror, a man with a definition written with blood by their Founding Fathers 233 years ago in the world's most brilliant political document. The second choice is easy and temporarily comfortable, but it is a choice reserved only for slaves, who, lacking courage to howl "give me death or freedom," must eventually lose chance to protest humiliation. This is a choice that will only be picked up by a selfish coward who has no concern about the suffering of his future descendants. He who picks up this choice deserves nothing but being conquered or even being wiped out.

Make no mistake, however, no matter how challenging the job of reviving America may be, no violent procedure of any extend should be allowed in this second revolution. With extraordinary intelligence and bravery, Americans completed the first revolution 233 years ago with muskets. Now, they must show to the world that they are capable of launching a second revolution with no less intelligence and bravery but with pens and will. Too many enemies of this country cannot wait to see a violent "revolution" to spread on this land. Let no American fall into this death trap.

Through the first revolution, Americans removed the dictatorship that shackled this land from top down; through the second revolution, Americans will remove the dictatorship that has been increasingly shackling this country up side down. Through the first revolution, Americans restructured the society with a government and principle that were unseen in human history.

Through the second revolution, Americans will refurbish and buttress the same social structure, same government, and same principle handed down by their Founding Fathers with a reshaped ideology of more fine tuned perfection at a new height.

As strenuous as the job of reviving America can be predicted, by the time the job is accomplished, a new generation of Americans would have towered over the world like their Founding Fathers with unmatched pride, a pride that the Founding Fathers expected their posterity to acquire and to sustain, a pride with which a people can tell the world that they have been nurtured by a best culture on a best land, a pride that is only reserved for and deserved by the citizens of the nation who can sing with their chins well risen: "Oh, the land of the free and the home of the brave."

To get to that generation, however, American people must win in the race against time.

Part IV

How Will *Relativity* Support Itself in Physics?

An award of $50,000, together with two smaller ones, is hereby posted by this author for a solution that shall liberate this author from the confusion in understanding some of the mathematical outcomes led by relativity. These outcomes are displayed at the end of the upcoming text that immediately follows this page. Limited by the education from the Newtonian school in physics, this author always feels that the principles demonstrated in relativity have created a realm not to allow him to enter. This award is the minimum that this author should offer to express his gratitude to a contributor who would submit a qualified solution. Volunteers are also needed to supervise the fund for the proposed award so that the fund will be independent of the control of this author, and fairness can then be confidently realized.

(The publisher of this book has nothing to do with this award offering and will not be responsible to this award in any part in any manner.)

1. Text

How will *Relativity* Support Itself in Physics?
(Part IV of the book *Aqua Soil*)

By *Cameron Rebigsol*

Frames AB and A'B' in the following diagram are passing each other at constant speed on each of their own straight lines, which are parallel to each other. An observer records the movement with a single clock while riding on AB and being stationary with respect to AB.

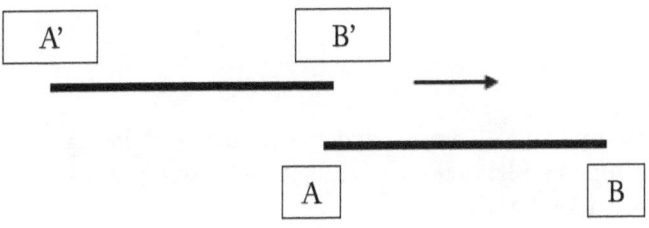

Fig. 1

As shown in figure 1, at time instant $t_0=0$, point A and point B' match each other in his observation. At some later time instant t_1, as shown in figure 2, B' is found matching point M in frame AB while A is found matching point M' in frame A'B'.

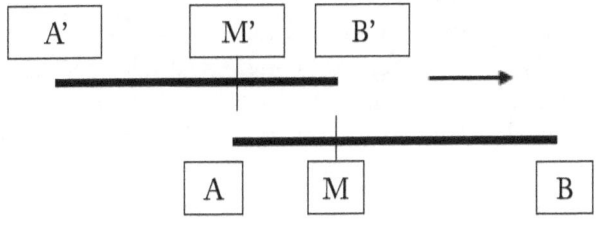

Fig. 2

So, with these recordings found in his laboratory, this observer expresses the speeds in the following ways:

$$v_1 = \frac{|AM|}{t_1}$$, a speed for A'B' moving with respect to frame AB, or the observer himself, and $v_2 = \frac{|B'M'|}{t_1}$, a speed for AB moving with respect to frame

A'B', where $|AM|$ and $|B'M'|$ are absolute lengths. If the spatial coordinate

values at A and B' are zero, naturally, the numerical reading for the coordinates at M and M' determines and expresses the speed value of each frame respectively. Therefore, we have

$$v_1 = \frac{|M|}{t_1} \text{ and } v_2 = \frac{|M'|}{t_1}$$

Since these two speeds are concluded with ruler and clock, through physical experiment done in a laboratory, we called these two speeds laboratory speed and further denote them as

$$v_1^{lab} = \frac{|M|}{t_1} \text{ and } v_2^{lab} = \frac{|M'|}{t_1}$$

The superscript "lab" here is only for denoting and has no meaning in mathematical operation. Speeds so experimentally concluded must have indisputable legitimacy in all schools of physics. What is left in dispute is whether or not the numerical values of $|M|$ and $|M'|$ are to be found equal. In other words, if a value of 3, for example, is read from M when B' matches it, will M' also show 3 to match point A or another value such as 4?

Newtonian physics believes that the numerical value $|M|$ and $|M'|$ of the spatial coordinates at M and M' must be absolutely identical to each other either through experimental measurement as mentioned above or through mathematical prediction that is concluded from other previous similar experiences. This must lead the observer to have

$$v_1 = v_2$$

Therefore, the Newtonian school will not consider movement to have different impact on the reading of $|M|$ and $|M'|$.

Relativity has no reason to decline the legitimacy on methods of concluding speeds through experiment as mentioned above. "If we wish to describe the *motion* of a material point," said A. Einstein, "we give values of its co-ordinates as functions of the time" (On the Electrodynamics of Moving Bodies, 1905). However, with its own reasoning, relativity does challenge the above conclusions drawn by the Newtonian school. With its concept of contraction of a moving length observed by a stationary observer, $|M'| = |M|$ may no longer hold; following it may also be a consequence of $v_1 \neq v_2$ because of the acceptance of

$$v_1^{lab} = \frac{|M|}{t_1} \text{ and } v_2^{lab} = \frac{|M'|}{t_1}.$$

In other words, relativity does not only accept the coordinate value shown on the point being matched—i.e., M' on A'B' as a legitimate value for determining and expressing a speed that is $v_2^{lab} = \frac{|M'|}{t_1}$—but also claims that the matched point must have a value that is different from $|M|$. If not different, the spatial coordinates contained by $|B'M'|$ will be precisely identical to that of $|AM|$, and denouncement of the property of length contraction asserted by relativity must follow. Subsequently, relativity would have placed itself inside the Newtonian school.

According to relativity's concept of length contraction, if a stationary length L from A'B' containing the same spatial coordinates as $|AM|$ is found moving by an observer on AB, the movement of L must make it be observed as having a shorter length, covering a collection of spatial coordinates on AB as

$$L_{moving} = |AM| \cdot \sqrt{1 - \left(\frac{v}{C}\right)^2},$$

where v is the value of speed between two moving frames that interests the observer like what we have been discussing, and C is the speed of light.

Based on this idea, if B'M' starts with a stationary length equals to $|AM|$, and by the time M' found its matching point A, B' will not coincide with point M but at a point whose value is determined by

$$|AM| \cdot \sqrt{1 - \left(\frac{v}{C}\right)^2}$$

Apparently, this is a value less than $|AM|$ unless $v=0$.

So, in order that B' coincides with point M at the completion of the time interval determined by t_1, M' must have a spatial coordinate value different from that of point M to start with. The length contraction concept thus must give this value as

$$|B'M'| = \frac{|AM|}{\sqrt{1-\left(\frac{v}{c}\right)^2}} \quad (Eq.\ 1)$$

Dividing both sides by t_1, we can have

$$\frac{|B'M'|}{t_1} = \frac{|AM|}{t_1 \cdot \sqrt{1-\left(\frac{v}{c}\right)^2}}$$

$$v_2^{lab} = \frac{v_1^{lab}}{\sqrt{1-\left(\frac{v}{c}\right)^2}} \quad (Eq.2)$$

Some consequences of *Eq.2* are the following:

1. It immediately tells people that speed is, in turn, a dependent quantity on another speed besides being impacted by speed of light.
2. Since both v_1^{lab} and v_2^{lab} are equally legitimate to be used in the place of v in relativity, v_1^{lab} and v_2^{lab} outside of the square root must be affected accordingly, depending on whether v_1^{lab} or v_2^{lab} is used inside the square root. Relativity thus allows no finite speed to be determined in a system that includes only two frames moving in relation to each other for the same observer staying in the same frame.
3. Upon matching, the numerical value of $|M'|$ that is shown opposite to point A cannot be uniquely found; but relativity must enable two of them: (1) the $|M'|$ value that permanently remains as an unaltered numerical figure for the stationary length of $|M'B'|$ to have been quoted, even after movement starts, and (2) a $|M'|$ value that must

have satisfied $\dfrac{|M'B'|\sqrt{1-\left(\frac{v}{c}\right)^2}}{t_1} = \dfrac{|AM|}{t_1}$ and is thus a converted value, then $|M'|$ must be a value depending on v. A more numerical illustration about this confusion is like this: According to (1), a value 4, for example, for M' can never change but is "deadly" attached to the length $|M'B'|$ at where point M' is marked, no matter how movement has changed. As to (2), If we mark the same point with an alphabetical sign of M' and a numerical figure 4 for the corresponding stationary length before the movement starts, at the time A and M' match, should the observer see M' remains unchanged as an alphabetical mark but the numerical figure of 4 would have changed to 3 to be justified for a certain non-zero speed?

Let us explore some outcome of *Eq.2* with some numerical figures.

If $v_1^{lab} = 0.6C$, using $v_1^{lab} = 0.6C$ in place of v, we have $v_2^{lab} = 0.75C$. If $v_1^{lab} = 0.6C$, using v_2^{lab} as an unknown in place of v, solving *Eq.2* will lead v_2^{lab} to an imaginary number. Therefore, v_2^{lab} cannot be uniquely determined.

If $v_1^{lab} = 0.8C$, using $v_1^{lab} = 08C$ in place of v, we have $v_2^{lab} = 1.33C$, a speed that relativity forbids to appear in nature with its principles. If $v_1^{lab} = 0.8C$, using v_2^{lab} as an unknown in place of v, solving *Eq.2* will lead v_2^{lab} to another imaginary number. Again, v_2^{lab} cannot be uniquely determined.

If we rearrange *Eq.2* to appear as

$$v_2^{lab} \sqrt{1 - \left(\frac{v}{C} \right)^2} = v_1^{lab}$$

Allowing $v_1^{lab} = C$, using $v = v_1^{lab} = C$, we will have $v_2^{lab} \cdot 0 = C$. Anything, even an infinitive number, times zero must end up as zero. So we can lead us to $0 = C$ with relativity.

Allowing $v_2^{lab} = C$, using $v_2^{lab} = C$ in place of v, the above expression leads to $v_1^{lab} = 0$. Does it mean that an observer seeing nothing moving ($v_1^{lab} = 0$) is a consequence of his movement of extraordinary speed with respect to something? Suppose we start with a stationary length of $|AM| = 0.6$ light-second with $t_1 = 1$ second. Definition of v_1^{lab} and *Eq. 1* will give us $|B'M'| = 0.75$ light-second as its stationary length. If the point value of $|M'|$ remains unchanged (why should it change?) from zero movement to the time it matches point A, definition of

$v_2{}^{lab}$ can give us a new value for v in *Eq. 1*. This inevitably leads to a new value of $|AM|$, which will contradict the initial value of 0.6 light-second.

All these numerical results are certainly mysteries. Obviously, principles from relativity and some fundamental mathematical rules cannot tolerate each other.

Besides mathematical difficulties, as a theory dealing with physical phenomenon, relativity has a void regarding how energy in the material world varies with the concept of length contraction. Coming back to the physical world, every mathematical coordinate system must attach to some material objects. So, if frame A'B' in the above discussion is marked on a glass rod or an ice rod or a steel rod, when each of them shows the same speed passing the observer on AB, will the rod of each material dissipate or generate different amount of energy for the same length contraction? How will also the thickness of each rod of the same material affect the energy variation due to length contraction?

2. Details for the $50,000 Award

Anyone who is interested in offering a solution to explain the confusion baffling this author can send the paper to

Mr. Cameron Rebigsol
11124 NE Halsey, PMB 577
Portland, OR 97220

Text of qualified paper should all be printed in English and contain no more than ten pages of 8x11 paper, single sided, double line, with letters printed in font size no smaller than 12.

No qualification of the solution provider is restricted. Each paper must be endorsed by a total of five professors with tenure. The endorsing professors should be still active in academic research or teaching in physics or mathematics or in electrical, mechanical, and civil engineering, and are expected to be still academically active for at least another two years after signing endorsement. For each endorsement, there must be an accompaniment of at least fifty signatures of students along with the paper. The signatures are from students majoring in mathematics, physics, electrical engineering, or mechanical or civil engineering in the universities. When signing, each student must have received one duplicate copy of the cover page of part 4 of the book *Aqua Soil* as well as the entire text portion following it. (Copyright of this part is hereby released for purpose of nonprofit duplication.) Each signature must be accompanied with verifiable contact information, such as telephone number, e-mail address, or mailing address. Please refer to the form on the last page of part 4.

Paper received without qualified solution will not be answered or returned unless an envelope with enough postage and return address is sent along with the paper. Award winner's name with contact information will be used to answer questions regarding award status from any concerned party. At this author's option, he may or may not publish any paper sent in, in spite of the winning status, with all names and contact information included in the paper, in any publishing media, but most probably in the website http://www.aquasoil.net.

Solution to be qualified as winning must include proof of at least the following:

Among the multiple speeds that are led by relativity and shown in the preceding text, only one in the system is legitimate.

Only a single value for M' is allowed by relativity, and this value can serve as an experimentally quoted value as well as a value to satisfy relativity's mathematical prediction.

In case counterargument sent in for award allows all multiple speeds to be equally legitimate, the same argument must prove that relativity will not lead to $0=C$, either $v>C$ as shown in the preceding text. Any argument that has two speeds or more involved, a mathematical expression showing the mathematical relationship between all multiple speeds must be provided.

If fortune can lead this author to have seven volunteers to supervise the fund for awarding, this author will send the total sum for award to a reputable but cost-free institution, which will be recommended by these volunteers and will be serving as an escrow office. Once money is sent there, this author will not be able to regain access to this money. It will be distributed in the following manners:

Pay the winners, but if winners finally do not appear, then the money will be equally divided and distributed among the volunteers if certain conditions are met. If the conditions cannot be met while no paper is qualified for winning, the money will be donated to some scholarship foundation at the recommendation of these volunteers for students majoring in physics and mathematics and engineering.

The last day to accept paper for solution is January 1, 2014. Since it cannot be known in advance when this author can have enough volunteers, it is recommended to each solution contributor to send a self-stamped envelope with return address along with the paper. In this way, when enough volunteers have committed themselves for the supervising job, the solution contributor will be notified whom and where to contact regarding the paper and the award status. Another way to contact this author is through the following e-mail address: <<*crebigsol@aquasoil.net*>>. (No paper will be accepted via e-mail.)

All volunteers must be professors with tenure from the universities in the list following this award announcement details. Professors endorsing paper for award will not be accepted as volunteers to supervise the award fund. Paper that is endorsed by professors who volunteer to supervise the award fund will be declined to be reviewed. The professors who volunteer must be currently academically active and expected to be academically active for at least another four years after being accepted as volunteers. Faith toward relativity with a positive and negative view is not a requirement to be accepted as a volunteer at all. However, when a volunteer is accepted, he/she must provide at least two hundred student signatures. Each signature must be accompanied with verifiable contact information, such as telephone number, e-mail address, or mailing address. Please refer to the form on the last page of part 4.

The total sum for the award fund for the winners is $52,500 USD. They will be given in the following manner:

According to the post day of the paper that is sent in, the earliest one who is qualified as winner receives $50,000, the second one receives $2,000, and the third one receives $500. If more than one paper is sent in on the same day and win, the winning papers of the same day will share the award for that date. For example, if the $50,000 award has been claimed, then three subsequent winning papers are found on a later day, these three papers of the same day will share the $2,000 award.

By March 31, 2014, if no winning paper is found, the money will be equally distributed to the volunteers, but each of the volunteers must also send to this author two written notes for the money to be released. Note one is a letter of recommendation to the university employing him/her that relativity is found with inexplicable mathematical contradiction and that students in related courses must be made aware of such contradiction. An acknowledgment (not necessarily an agreement) of such recommendation from the university must also be accompanied. Note two is to state that in all his/her future lectures, if relativity must be mentioned, the volunteering professor will make students aware of the mathematical difficulties that relativity should overcome before it can be accepted as a positive theory. Without compensation, all two notes may be published by this author in any manner with any publishing media, most likely in the website http://www.aquasoil.net. If these two notes cannot be obtained from every and all the volunteers by this author, then the entire fund will be transferred to the appropriate institution and be used as scholarship fund.

This author is fully aware that $52,500 is far from being enough to compensate the effort that the volunteers will contribute. However, this author also believes that their nobility they have been holding in the long journey of pursuing truth will make them balance their work with the value of truth, not the value of money.

When finally no winner is found and if, besides sending in the aforementioned two notes to this author, the volunteering professors would enable the preceding text to be published in journals like the *Nature* or *Journals of the American Physical Society* or equivalent media, this author will increase the compensation to a total of $10,000 USD, including the equally divided distribution from the $52,500, to each of them, and this author will pay all the cost of publishing.

3. List of Universities

Papers endorsed by professors with tenure from all of the following universities or colleges will be accepted for award reviewing. This author will feel tremendously honored if professors with tenure from the university with "*" would like to volunteer to supervise and review the corresponding papers. In total, seven volunteers are needed. If a professor is interested, please contact this author at http://www.aquasoil.net. For the time being, only professors from the universities with "*" are needed.

USA		Alfred University
USA		American University
USA		Amherst College
USA		Bard College
USA		Bates College
USA		Bentley University
USA	*	Boston University
USA		Bowdoin College
USA		Brandeis University
USA		Brigham Young University
USA	*	Brown University
USA		Bryn Mawr college
USA		Bucknell University
USA		Cal Poly
USA	*	California Institute of Technology
USA	*	Carnegie Mellon University
USA		Claremont McKenna University
USA		Clark University
USA		Colgate University
USA		College of Charleston
USA		College of William and Mary
USA		Colorado College
USA	*	Columbia University
USA		Connecticut College
USA	*	Cornell University
USA	*	Dartmouth College
USA		Davidson College
USA	*	Duke University
USA	*	Emory University
USA		George Washington University

USA		Georgetown University
USA		Georgia institute of Technology
USA		Gettysburg College
USA		Grinnell College
USA		Hamilton College
USA	*	Harvard University
USA		Harvey Mudd College
USA		Haverford College
USA		Howard University
USA		Indiana University
USA		James Madison University
USA	*	Johns Hopkins University
USA		Kenyon College
USA		Lafayette College
USA		Lawrence University
USA		Macalester College
USA		McGill University
USA		Middlebury College
USA	*	Massachusetts Institute of Technology
USA	*	New York University
USA		Northeastern University
USA	*	Northwestern University
USA		Occidental College
USA		Ohio State University
USA		Ohio University
USA		Pomona College
USA	*	Princeton University
USA		Providence College
USA	*	Purdue University
USA		Reed College
USA		Rice University
USA		Rochester Institute of Technology
USA		Saint Joseph's University
USA		Scripps College
USA		Smith College
USA		St. Olaf College
USA	*	Stanford University
USA		Swarthmore College
USA		Trinity University
USA		Trinity University—Texas
USA		Tuft University

USA		University of California, Santa Barbara
USA	*	University of California, Berkeley
USA		University of California, Davis
USA		University of California, Irvine
USA	*	University of California, Los Angeles
USA		University of California, Riverside
USA	*	University of California, San Diego
USA		University of California, Santa Cruz
USA		University of Arizona
USA	*	University of Chicago
USA		University of Cincinnati
USA		University of Connecticut
USA		University of Delaware
USA		University of Denver
USA		University of Florida
USA		University of Georgia
USA	*	University of Illinois
USA		University of Iowa
USA		University of Kansas
USA	*	University of Maryland
USA		University of Miami
USA	*	University of Michigan
USA		University of Minnesota
USA		University of North Carolina
USA		University of Notre Dame
USA		University of Oklahoma
USA		University of Oregon
USA	*	University of Pennsylvania
USA	*	University of Pittsburgh
USA		University of Rhode Island
USA		University of Richmond
USA		University of Rochester
USA		University of South California
USA		University of South Carolina
USA	*	University of Texas at Austin
USA		University of Virginia
USA	*	University of Washington
USA		University of Wisconsin-Madison
USA		Ursinus College
USA		Vanderbilt University
USA		Vassar College

USA		Virginia Tech
USA		Washington & Lee University
USA		Washington University in St. Louis
USA		Wellesley College
USA		Wesleyan University
USA		West Point Military Academy
USA		William College
USA	*	Yale University
Argentina		University of Buenos Aires
Australia		Australian National University
Brazil		University of Sao Paulo
Canada	*	University of Toronto
Canada	*	The University of British Columbia
Canada		University of Waterloo
Canada		York University
Canada		McGill University
China	*	Peking University
China	*	Tsinghua University
France		University of Paris 06
Germany	*	University of Munich
Germany		Technical University of Munich
Greece		University of Athens
Hong Kong		University of Hong Kong
India		University of Delhi
Israel		Hebrew University, Jerusalem
Italia		University of Rome La Sapienza
Japan	*	Tokyo University
Japan		Kyoto University
Japan		Tokyo Institute of Technology
Russia	*	Lomonosov Moscow State University
Russia		St. Petersburg State University
Spain		University of Barcelona
Taiwan		National Taiwan University
UK	*	University of Oxford
UK	*	University of Cambridge
UK		University of Edinburgh
UK		University of Manchester

4. Forms for Signature Collection

When collecting student signatures, forms like the two ones following this page should be used. The form for endorsing a paper and the form for volunteering professors are slightly different.

Signature Collection for Endorsement of Solution Paper

If a student is asked to enter his/her name in this form, he/she must be given a free copy of 'How Will Relativity Support Itself in Physics" with the $50,000 award announcement from Part IV of the book "*Aqua Soil*' by Cameron Rebigsol, published by Xlibris Coperation.

Professor's Name: _____ **University/College :** _____

Contact Information _____

Country: _____

| Student's Name | Contact Information (Fill out any two) | | | Major | Year of Graduation | Today's date |
	Mailing Address	Email Address	Telephone #			

Signature Collection for a Volunteering Professor

If a student is asked to enter his/her name in this form, he/she must be given a free copy of "How Will Relativity Support Itself in Physics" with the $50,000 award announcement from Part IV of the book "*Aqua Soil*" by Cameron Rebigsol, published by Xlibris Coperation.

Professor's Name: _____ **University/College :** _____

Contact Information _____

Country: _____

Escrow office recommanded: _____

| Student's Name | Contact Information (Fill out any two) | | | Major | Year of Graduation | Today's date |
	Mailing Address	Email Address	Telephone #			

References

The Declaration of Independence of America

The Articles of Confederation

The American Constitution

The Bill of rights

Smithsonian Intimate Guide to Human Origins, by Carl Zimmer, HarperCollins Publishers Inc.

The Encyclopedia of Mammals, edited by Prof. David MacDonald, copy right © 2001 the Brown Reference Group

Wild Things life as we know it, by Amanda Bensen, Jess Blumberg, T.A. Frail, Megan Gambino and Laura Helmuth, *Smithsonian* magazine, Feb. 2008

The Rise of Mammals, by Rick Gore, *National Geographic* magazine, April 2003

New Find, By Rick Gore, *National Geographic* magazine, Aug. 2002

What on Earth is Plate Tectonics? US Geological Survey, National Park Services, Department of the Interior

From Ape to Adam, By Herbert Wendt, The Bobbs-Merrill Company, Inc. Indianapolis, New York © 1972 Thames and Hudson Ltd, London

Lovers, Not Fighters? New genetic signs that modern humans mated with Homo erectus, By John Whitfield, *Scientific American* magazine, March, 2008

Race, Brain Size, and IQ, by the General Psychologist,

http://www.apa.org/divisions/div1/news/summer2002/rushtonpdf.pdf, Nov 15, 2007

Living Tribes, by Colin Prior, Carolyn Fry, Firefly Books (U.S.) Inc.

Human, DK Publishing, Inc., Copyright © 2004, 2006 Dorling Kindersley Limited

State of Emergency, by Patrick J. Buchanan, Thomas Dunne Books.

Common Sense, by Glenn Beck, Mercury Radio Arts/Threshold Editions.

Chronicle of the Roman Republic, by Philip Matyszak, Thames & Hudson.

British History for Dummies, by Sean Lang, John Wiley & Sons, Ltd

European History for Dummies, by Sean Lang, John Wiley & Sons, Ltd

The Bicentennial Almanac, 200 Years of America, by Calvin D. Linton, Thomas Nelson, Inc., Publishers

Atlas of Jewish Civilization, by Martin Gilbert and Josephine Bacon, Quantum Books Ltd

Mao, *The Unknown Story*, by Jung Chang and Jon Halliday, a Borzoi Book published by Alfred A Knopf

Relativity, *the special and the General Theory*, written by Albert Einstein, translated by Robert W. Lawson, Three Rivers Press

Glossary

aqua ape: An amphibious primate that is conceived as having its tail lost in evolution and spending its major life in an aqua environment. Although it is reasoned as having a posture of a low sinking chin toward its chest like arboreal apes, it retains a limb proportion of longer legs than arms. Compared to other primates, its body has the highest content of body fat.

aqua hominid: A highly intelligent primate that is still dwelling primarily in water. Aqua hominid can be conceived as the direct ancestor of terrestrial hominid, i.e., the hominid that we have been always referred to in human evolution before this new term is introduced.

face plane: A geometrical plane that is defined by the following three points: tip of the snout of an animal and its two eye balls. If we further draw on the plane an arrow that starts from the snout and points toward the center between the two eyes, then a plane is positive if it points away from the animal's rear end, and is negative if it points toward its rear end. A less simplified, or more restricted, definition of the face plane of a human being should be established in the following way: Let us find on a person's skin near the nose ridge a point, called A, which is the center between the two eyebrows. We further connect this point and his central tip of the upper lip with a straight line called AB. Let us also have a line, called CD, connecting the center of the two pupils. The human's face plane will be a plane that is through line AB but parallel with CD. (Line CD may or may not lie in the face plane)

primitive primate: It is supposed to be the oldest version of primate from which all lineages of primate should be derived from.

primitive aqua primate: Primitive primates that dwells in water.

Index

life span of, 157-65
Deng Xiao-Ping, 176
developments, physical, 49, 52
dictatorship, 88, 106, 141, 158-59, 163-
    65, 175-76, 178-80, 203, 262
  syndrome of, 163
DNA, 12, 55, 57, 61, 75, 147

E

ECHR (European Convention on Human
    Rights), 112-13, 116, 118-19
education, 92, 100, 104-5, 129, 170-71
energy reservoir, two openings of
  obligation input, 167-71, 245
  right outlet, 167-68
Engels, Friedrich, 178
English, nation of, 195-97
European Convention on Human
    Rights, 112-13, 116, 118-19
evolution
  biological
    different lineages, 42
    fundamental characteristic
      developments, 17, 24
    human races, 81
    origin, 43, 62
  racial
    different concern, 126
    hatred, 144
    new racism, 127-34
eyes, 67-72

F

face, 64-65
face plane, 22, 40, 50, 54, 67, 71
facial feature, no change on the, 73
fairness, fatality of
  concept of fairness, 98-100
  modern social effect of fairness, 101-3

ultimate fairness, ultimate
    monopolization, 103-7
Far East Asia, 65, 71-73, 77, 148
feet, different, 27, 29
fire, 78-79, 192, 222
forelimbs, advancing of, 22
Founding Fathers, 189-94, 196-97, 225-28,
    231-32, 245, 247-48, 261, 263-64

G

geladas, 39, 58
gene copies, box of, 59-61
genetic selection, 76, 78
Genghis Khan (Mongol conqueror), 135
Germany, 17, 57, 139-40, 279
God, 11, 86, 191, 197, 220, 224, 226
Goering, Hermann, 140
golden rules, 174-79
gorilla, 9, 18, 28, 31, 36, 42, 49, 52, 62
grave digging, 88, 92
Great Britain, 139
Gypsy, 136

H

habitats, varying, 26
Hagia Sophia, 147
hair form, 77. *See also* scalp hair
hatred
  new, 140-41
  who needs, 142-44
Hitler, Adolph, 138
Holocaust, 140
hominids, 9, 13, 40-41, 52-53, 56-57,
    59, 61-63, 66, 73-75, 77-80
  aqua, 9, 40-41, 52, 56, 78
  Homo erectus, 56, 228
  Homo ergaster, 9, 41, 52, 126
  Homo sapiens, 28, 52-53, 56-57, 62-63,
    65, 73, 85, 125

288

www.ingramcontent.com/pod-product-compliance
Lightning Source LLC
Chambersburg PA
CBHW031825170526
45157CB00001B/192